BIOMECHANICS
BASIC FUNDAMENTALS

BIOMECHANICS
BASIC FUNDAMENTALS

by
Jinesh K B
Anindya Saha
R R Sharma

Wordpen Academics
Publishing Science That Matter

ISBN: 978-93-6084-457-8

Published by : **WORDPEN ACADEMICS**
4736/23, Ansari Road, Darya Ganj,
New Delhi – 110 002
Phone: +91-011-35004336
E-mail: info@wordpenacademics.com

Printed at **:** **Replika Press Pvt. Ltd.**

Exclusive Distributor: ***Bio-Green Books, New Delhi***

Preface

Biomechanics: Basic Fundamentals offers a comprehensive introduction to the essential principles and concepts underlying the field of biomechanics, focusing on the mechanics of biological systems and the application of physical principles to human movement and performance. This book is designed for students, researchers, and professionals in biomechanics, kinesiology, physical therapy, sports science, and related fields, providing a solid foundation for understanding both the theoretical aspects and practical applications of biomechanics in various domains.

Biomechanics is the study of how the laws of physics apply to living organisms, particularly human movement. It encompasses a wide range of topics, including the analysis of forces, motion, balance, and the mechanical properties of biological tissues. From understanding the mechanics of walking and running to analyzing the performance of athletes or designing prosthetics and orthotics, biomechanics plays a crucial role in enhancing human health, performance, and rehabilitation.

This book aims to present the fundamental concepts of biomechanics in a clear and structured manner, beginning with an exploration of basic physical principles such as force, work, energy, and motion. The core topics of kinematics, kinetics, muscle function, joint mechanics, and posture are then discussed in detail, providing readers with a well-rounded understanding of how these concepts are applied to the human body. Furthermore, the book highlights the tools and techniques used to measure and analyze biomechanical parameters, such as motion capture, force plates, and electromyography.

Throughout the book, readers will find practical examples, case studies, and applications of biomechanics in various fields, including sports, rehabilitation, ergonomics, and injury prevention. Special attention is given to the evolving role of biomechanics in advancing technology, such as the development of wearable devices, assistive technologies, and computational modeling techniques.

The contributions from leading experts in the field ensure that Biomechanics: Basic Fundamentals is both scientifically rigorous and accessible, offering readers a solid theoretical grounding while also emphasizing the practical implications of biomechanics in real-world settings. The book aims to equip students with the knowledge and skills necessary to apply biomechanics to improve human movement, performance, and health outcomes.

We are grateful to the authors and contributors who have shared their expertise to make this book a valuable resource. It is our hope that Biomechanics: Basic Fundamentals will serve as a foundation for further exploration and innovation in the dynamic field of biomechanics.

Jinesh K B

Anindya Saha

R R Sharma

Contents

Chapter 1
Introduction

1.1 Why Study Biomechanics?

Although physical educators love to move, studying movement from a biomechanical perspective is not something many teachers get excited about, subsequently neither do their students. The subject matter is often considered hard and complex and this is probably justified given the equations and calculations many of us were presented with when studying the topic. It does however, have the potential to be one of the most interesting areas of study within Physical Education. Subjects like anatomy, physiology, history, health and the others, while all essential, and are not as responsive to an intervention as biomechanics. For example, have someone run between cones with a ball held in two hands and then again without a ball and you will instantly see different movement strategies being used. This facet of biomechanics can be used as a tool to help you present concepts more effectively.

Now that we have seen it has potential, what actually is Biomechanics? We use this term, and many others, such as Kinesiology, technique analysis, movement analysis or skill analysis all the time, but what do they really mean? Kinesiology is the scholarly study of human movement and biomechanics is one of its academic sub disciplines. To find the root of the word 'biomechanics' we need to go back in time. The term 'mechanics' is used as a description for of a branch of physics that is concerned with the motion of objects that are acted on by forces. The concept dates back to Archimedes although more recent major contributions were from Galileo in the 16th century and Newton in the 17th century. Bio-mechanics combines the principles of mechanics with the fields of biology and physiology and can be traced back to Leonardo da Vinci in the 15th century. Two common areas within biomechanics are Kinematics, which is used to describe motion and Kinetics, which considers the forces and masses in the analysis.

Now that we know what it is we are talking about, why should we study it? There are many reasons to study movement and these range from community health through to elite sport or just for the sheer joy of studying the human body. With regard to community health, there are an increasing number of initiatives to get people active and moving so it is important that when they move they move correctly. Considering correct technique in the overall health plan can perhaps increase enjoyment and participation and also reduce a potential abundance of overuse injuries that may occur as an inactive population gets moving.

Another reason to study movement, relates a desire to improve performance. We often watch people undertaking complex movements in dance or sports and talk about the skilled performer having better form or 'flow' than a lesser skilled performer. However, when it comes to a biomechanical perspective we are really talking about the pattern or sequence of movements someone uses to perform a skill. How good someone will get at a skill depends on both their genetic endowment and what they learn, how much influence each has on the overall performance is unknown. What we do know is that we are all given a genetic window of potential and physical educators can have a huge impact as to where the student lies within that window, therefore it is important that they have a good understanding of the movement sciences.

We can also be interested in movement analysis just because we appreciate the human body.

Physical educators may be interested in movements ranging from an intricate dance piece to leisure walking. Perhaps too often we underestimate just how amazing our bodies actually are.

The best engineers in the world have only just managed to create a robot to 'look and walk' at a very basic level, to have it hold a conversation and chew gum at the same time is far beyond its capabilities, something the human body does almost automatically. The human body is the most complex thing that any of us will ever operate; it comes with no user manual but only our innate gene coded potential and what we learn. We can't easily change our genes so our best options to improve performance are teaching and learning.

1.2 What Is Biomechanics?

Biomechanics is the study of structure, forces, and the effects of forces on structure. So what does that have to do with exercise and sport? Every exercise in which you participate, every movement, every effort to prevent motion, and every physical challenge involves your human structure and very real forces.

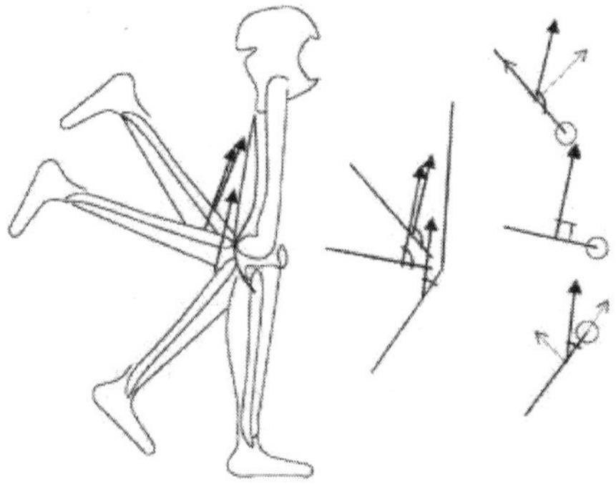

Fig structure and forces

Pursuing a deeper understanding of human structure and function, external forces applied to the structure, and the internal forces and responses that occur as a result, are what will give us real and solid perspective from which to analyze and progress exercise. Viewing exercise from a biomechanical perspective will help us to become immune to much of the pseudoscience, gym myths, and fallacies that continue to mislead the vast majority of the sports and exercise world.

Classified as both a basic and applied science, biomechanics integrates the laws of physics with the working concepts of engineering to describe the internal and external forces acting on the body as well as the effects created with these forces. Although biomechanics is more focused on the mechanical principles of body systems, there are many times when physiological functions and biochemical tissues are taking into consideration when discussing motion. Studies in biomechanics often will address a broad range of topics, including the mechanical function of muscles, connective tissues, cartilage, skin, nerves, joints, bones, and even internal organs. Biomechanics is an increasingly important area of scientific study for addressing problems related to human health and performance. Biomechanics research is often applied by physical education teachers, physical therapists, physicians, personal trainers, and coaches.

Biomechanics combines the field of engineering mechanics with the fields of biology and physiology. Biomechanics applies mechanical principles to the human body in order to understand the mechanical influences on bone and joint health. Forces that load the joints are generated by muscles and transmitted by tendons. Bones must withstand these forces. Developments in the field of biomechanics have improved our understanding of normal and pathologic gait, mechanics of neuromuscular control, and mechanics of growth and form. This knowledge has contributed to the development of medical diagnostic and treatment procedures. It has provided the basis for the design and manufacture of medical implants and orthotic devices and has enhanced rehabilitation therapy practices. Biomechanics has also been used to improve human performance in the workplace and in athletic competition.

One of the greatest limitations in biomechanics is that we cannot, with current technology, measure muscle forces in a noninvasive way. However, we can measure the acceleration of the limbs and forces between the body and the ground to calculate the net joint moment (NJM), which is the moment required to accelerate a limb in accordance with Newton's second law. Despite the fact that muscles and other soft tissue structures contribute to the NJM, and co contractions of the antagonists can make the actual moment much greater than the NJM, we usually equate high NJMs with high muscle forces needed to produce that moment. So when a research study suggests that exercise A has a greater extensor NJM at the knee than exercise B, it assumes that there is no co-contraction of the hamstrings during both exercises, and exercise A has a higher demand on the quadriceps. Studies will normally report internal moments (as described previously) or external moments (which are as a result of external forces and inertia). Internal moments are equal in magnitude and opposite in direction to the external moment.

Biomechanics is the application of mechanical engineering principles to living organisms. Biomechanics of organisms can be examined at different levels: cellular level (e.g., response of cells to an externally applied force or deformation), tissue level (e.g., strain of the anterior cruciate ligament during normal gait), and whole-joint level (e.g., joint contact forces during activities of daily living).

Engineering principles may be used to understand the cause and progression of many rheumatic diseases. A basic understanding of these principles is beneficial for clinicians and medical professionals. This chapter presents a rudimentary background of engineering mechanics pertaining to whole-joint and tissue mechanics in the human body.

1.2.1 Features

Biomechanics is closely related to engineering, because it often uses traditional engineering sciences to analyze biological systems. Some simple applications of Newtonian mechanics and/or materials sciences can supply correct approximations to the mechanics of many biological systems. Applied mechanics, most notably mechanical engineering disciplines such as continuum mechanics, mechanism analysis, structural analysis, kinematics and dynamics play prominent roles in the study of biomechanics.

Usually biological systems are much more complex than man-built systems. Numerical methods are hence applied in almost every biomechanical study. Research is done in an iterative process of hypothesis and verification, including several steps of modeling, computer simulation and experimental measurements.

Biomechanics is considered one of the core disciplines to understand principles of physiology and pathophysiology in the human. Decades of biomechanics research revealed many primary principles regulating growth, tissue formation, organism morphogenesis, and tissue regeneration. Not surprisingly, biomechanics is one of the core sciences used to understand how dynamic loading in the central nervous system (CNS) causes injury. This key transition from physiology to pathophysiology is especially important, as early pathophysiologic changes to the CNS can strongly influence cellular decisions to survive, reintegrate, and repair the injured CNS. This long-term response, initiated when forces are transferred to the cellular/molecular scale at the time of injury, is a key contributor in the outcome of head-injured and spinal cord-injured patients.

Compared to nearly every other application of biomechanics and human body, biomechanics of traumatic loading has three unique components.

- First, the mechanical event is nearly always considered as a single event, rather than a series of cyclical loadings that are applied to organ/tissue/cellular preparations. In other areas of biomechanics, cyclical loading of tissue/cellular components is critical in defining the homeostatic response and the adaptation of this response during disease. Examples include the periodic distension of the vascular wall during the cardiovascular cycle, or the frequent loading and unloading of orthopedic soft tissues during gait. In contrast, the brain and spinal cord are considered "mechanically protected" organs and do not have a clear constant level of mechanical stimulation.
- Second, traumatic loading is probably the fastest event studied in biomechanics, especially considering the very recent work on blast injury biomechanics. The very brief nature of the event – the acceleration/deceleration event in a blunt impact event is typically delivered in less than 50 milliseconds, while the mechanical loading in a blast event lasts only a few milliseconds – present enormous technical challenges because traditional biomechanical tools to study the forces, displacements, deformations, and stresses in tissue/cellular structures are often lacking.
- Third, the functional and physical wiring of the brain and spinal cord remain largely a mystery and are the focus of many efforts across the neuroscience and neurologic sciences. Any biomechanically based study of events at the cellular and molecular scale must consider the physical wiring of the structure and our ability to mechanically interrogate individual components of these circuits to lay the groundwork for a more integrative view of how the brain and spinal cord will respond to the mechanical forces occurring during injury.

Given this unique set of circumstances, it is not surprising that there has been a large increase in the use and development of simplified laboratory models to deconstruct some of these experimental complexities. Foremost among these challenges, is, developing new simplified laboratory models that develop repeatable, reproducible, and informative models of the clinical condition. Certainly, as we know more about how mechanical loading affects the function of the brain and spinal cord at many length scales – at the organ level, at the tissue level, and at the cellular/molecular level – we will become much more facile with developing models that accurately reproduce these loadings in vitro.

The conversion of the mechanical input to the resulting biochemical cascades – termed mechano-transmission – triggers the evolving injury patterns in the brain and spinal cord after traumatic injury. The concept of mechano-transmission has transformed biomechanics from a study of structure-property relationships into a more integrated structure-function-property triad across lengths scales in the CNS. At the most reductionist level, this means that we are now beginning to explore how some cells respond directly to the mechanical load with a set of biochemical signatures, even now extending these into genomic signatures. Not surprisingly, as the mechanical load is adjusted to either individual cells or clusters of cells, one may find that some cell types respond while others do not. It is not meant to be an exhaustive review of all work generated to date. Several other recent reviews can be collected and reviewed in their own right for this purpose.

1.2.2 Fields:

Sports biomechanics

- In sports biomechanics, the laws of mechanics are applied to human movements in order to gain a greater understanding of athletic performance and to reduce sport injuries as well.
- It focuses on the application of the scientific principles of mechanical physics to understand movements of action of human bodies and sports implements such as cricket bat, hockey stick and javelin etc.
- Elements of mechanical engineering (e.g., strain gauges), electrical engineering (e.g., digital filtering), computer science (e.g., numerical methods), gait analysis (e.g., force platforms), and clinical neurophysiology (e.g., surface EMG) are common methods used in sports biomechanics.
- Biomechanics in sports can be stated as the muscular, joint and skeletal actions of the body during the execution of a given task, skill and/or technique.
- Proper understanding of biomechanics relating to sports skill has the greatest implications on: sport's performance, rehabilitation and injury prevention, along with sport mastery.
- As noted, one could say that best athlete is the one that executes his or her skill the best.

Continuum biomechanics

- The mechanical analysis of biomaterials and bio-fluids is usually carried forth with the concepts of continuum mechanics. This assumption breaks down when the length scales of interest approach the order of the micro structural details of the material.
- One of the most remarkable characteristic of biomaterials is their hierarchical structure. In other words, the mechanical characteristics of these materials rely on physical phenomena occurring in multiple levels, from the molecular all the way up to the tissue and organ levels.
- Biomaterials are classified in two groups, hard and soft tissues.
- Mechanical deformation of hard tissues (like wood, shell and bone) may be analyzed with the theory of linear elasticity.

- On the other hand, soft tissues (like skin, tendon, muscle and cartilage) usually undergo large deformations and thus, their analysis rely on the finite strain theory and computer simulations. The interest in continuum biomechanics is spurred by the need for realism in the development of medical simulation.

Bio-fluid mechanics

- Biological fluid mechanics, or bio-fluid mechanics, is the study of both gas and liquid fluid that flows in or around biological organisms. An often studied liquid bio-fluids problem is that of blood flow in the human cardiovascular system.
- Under certain mathematical circumstances, blood flow can be modelled by the Navier–Stokes equations.
- In vivo whole blood is assumed to be an incompressible Newtonian fluid. However, this assumption fails when considering forward flow within arterioles.
- At the microscopic scale, the effects of individual red blood cells become significant, and whole blood can no longer be modelled as a continuum. When the diameter of the blood vessel is just slightly larger than the diameter of the red blood cell the Fahraeus–Lindquist effect occurs and there is a decrease in wall shear stress.
- However, as the diameter of the blood vessel decreases further, the red blood cells have to squeeze through the vessel and often can only pass in single file. In this case, the inverse Fahraeus–Lindquist effect occurs and the wall shear stress increases.
- An example of a gaseous bio-fluids problem is that of human respiration. Recently, respiratory systems in insects have been studied for bio-inspiration for designing improved microfluidic devices.

Bio-tribology

- The main aspects of Contact mechanics and tribology are related to friction, wear and lubrication.
- When the two surfaces come in contact during motion i.e. rub against each other, friction, wear and lubrication effects are very important to analyze in order to determine the performance of the material.
- Bio-tribology is a study of friction, wear and lubrication of biological systems especially human joints such as hips and knees.
- Femoral and tibial components of knee implant routinely rub against each other during daily activity such as walking or stair climbing. If the performance of tibial component needs to be analyzed, the principles of biotribology are used to determine the wear performance of the implant and lubrication effects of synovial fluid.

- In addition, the theory of contact mechanics also becomes very important for wear analysis.
- Additional aspects of biotribology can also include analysis of subsurface damage resulting from two surfaces coming in contact during motion, i.e. rubbing against each other, such as in the evaluation of tissue engineered cartilage.

Comparative biomechanics

- Comparative biomechanics is the application of biomechanics to non-human organisms, whether used to gain greater insights into humans (as in physical anthropology) or into the functions, ecology and adaptations of the organisms themselves. Common areas of investigation are Animal locomotion and feeding, as these have strong connections to the organism's fitness and impose high mechanical demands. Animal locomotion, has many manifestations, including running, jumping and flying. Locomotion requires energy to overcome friction, drag, inertia, and gravity, though which factor predominates varies with environment.
- Comparative biomechanics overlaps strongly with many other fields, including ecology, neurobiology, developmental biology, ethology, and paleontology, to the extent of commonly publishing papers in the journals of these other fields. Comparative biomechanics is often applied in medicine (with regards to common model organisms such as mice and rats) as well as in biomimetics, which looks to nature for solutions to engineering problems.

Plant biomechanics

- The application of biomechanical principles to plants, plant organs and cells has developed into the subfield of plant biomechanics.
- Application of biomechanics for plants ranges from studying the resilience of crops to environmental stress to development and morphogenesis at cell and tissue scale, overlapping with mechanobiology.

Computational biomechanics

- Computational biomechanics is the application of engineering computational tools, such as the Finite element method to study the mechanics of biological systems. Computational models and simulations are used to predict the relationship between parameters that are otherwise challenging to test experimentally, or used to design more relevant experiments reducing the time and costs of experiments.
- Mechanical modeling using finite element analysis has been used to interpret the experimental observation of plant cell growth to understand how they differentiate, for instance.

- In medicine, over the past decade, the Finite element method has become an established alternative to in vivo surgical assessment.
- One of the main advantages of computational biomechanics lies in its ability to determine the endo-anatomical response of an anatomy, without being subject to ethical restrictions. This has led FE modeling to the point of becoming ubiquitous in several fields of Biomechanics while several projects have even adopted an open source philosophy (e.g. BioSpine).

1.2.3 Subfields

Applied subfields of biomechanics include:

- Soft body dynamics
- Kinesiology (kinetics + physiology)
- Animal locomotion & Gait analysis
- Musculoskeletal & orthopedic biomechanics
- Cardiovascular biomechanics
- Ergonomy
- Human factors engineering & occupational biomechanics
- Implant (medicine), Orthotics & Prosthesis
- Rehabilitation
- Sports biomechanics
- Allometry
- Injury biomechanics
- Biotribology
- Bio-fluid mechanics
- Comparative biomechanics
- Computational biomechanics

1.2.4 Renaissance

The next major biomechanic would not be around until 1452, with the birth of Leonardo da Vinci. Da Vinci was an artist and mechanic and engineer. He contributed to mechanics and military and civil engineering projects. He had a great understanding of science and mechanics and studied anatomy in the a mechanics context of mechanics. He analyzed muscle forces and movements and studied joint functions. These studies could be considered studies in the realm of biomechanics. Leonardo da Vinci studied anatomy in the context of mechanics. He analyzed muscle forces as acting along lines connecting origins and insertions, and studied joint function. Da Vinci tended to mimic some animal features in his machines. For example, he studied the flight of birds to find means by which humans could fly;

and because horses were the principal source of mechanical power in that time, he studied their muscular systems to design machines that would better benefit from the forces applied by this animal.

In 1543, Galen's work, On the Function of the Parts was challenged by Andreas Vesalius at the age of 29. Vesalius published his own work called, On the Structure of the Human Body. In this work, Vesalius corrected many errors made by Galen, which would not be globally accepted for many centuries. With the death of Copernicus came a new desire to understand and learn about the world around people and how it works. On his deathbed, he published his work, On the Revolutions of the Heavenly Spheres. This work not only revolutionized science and physics, but also the development of mechanics and later bio-mechanics.

Galileo Galilee, the father of mechanics and part time biomechanic was born 21 years after the death of Copernicus. Galileo spent many years in medical school and often questioned everything his professors taught. He found that the professors could not prove what they taught so he moved onto mathematics where everything had to be proven. Then, at the age of 25, he went to Pisa and taught mathematics. He was a very good lecturer and students would leave their other instructors to hear him speak, so he was forced to resign. He then became a professor at an even more prestigious school in Padua. His spirit and teachings would lead the world once again in the direction of science. Over his years of science, Galileo made a lot of biomechanical aspects known. For example, he discovered that "animals' masses increase disproportionately to their size, and their bones must consequently also disproportionately increase in girth, adapting to load bearing rather than mere size. The bending strength of a tubular structure such as a bone is increased relative to its weight by making it hollow and increasing its diameter. Marine animals can be larger than terrestrial animals because the water's buoyancy [sic] relieves their tissues of weight."

Galileo Galilei was interested in the strength of bones and suggested that bones are hollow because this affords maximum strength with minimum weight. He noted that animals' bone masses increased disproportionately to their size. Consequently, bones must also increase disproportionately in girth rather than mere size. This is because the bending strength of a tubular structure is much more efficient relative to its weight. Mason suggests that this insight was one of the first grasps of the principles of biological optimization.

In the 16th century, Descartes suggested a philosophic system whereby all living systems, including the human body (but not the soul), are simply machines ruled by the same mechanical laws, an idea that did much to promote and sustain biomechanical study. Giovanni Alfonso Borelli embraced this idea and studied walking, running, jumping, and the flight of birds, the swimming of fish, and even the piston action of the heart within a mechanical framework. He could determine the position of the human center of gravity, calculate and measured inspired and expired air volumes, and showed that inspiration is muscle-driven and expiration is due to tissue elasticity. Borelli was the first to understand that the levers of the musculoskeletal system magnify motion rather than force, so that muscles must produce much larger forces than those resisting the motion. Influenced by the work of Galileo, whom he personally knew, he had an intuitive understanding of static equilibrium in various joints of the human body well before Newton published the laws of motion.

1.2.5 Industrial era

The next major bio-mechanic, Giovanni Alfonso Borelli, was the first to understand that "the levers of the musculature system magnify motion rather than force, so that muscles must produce much larger forces than those resisting the motion". Using the works of Galileo and building off from them, Borelli figured out the forces required for equilibrium in various joints of the human body. He even discovered the human center of gravity and air volume as well as muscle elasticity. His work is often considered the most important in the history of bio-mechanics because he made so many new discoveries that opened the way for the future generations to continue his work and studies.

It was many years after Borelli before the field of bio-mechanics made any major leaps. After that time since then, more and more scientists took to learning about the human body and its functions. There are not many notable scientists from the 19th or 20th century in bio-mechanics because the field is far too vast now to attribute one thing to one person. However, the field is continuing to grow every year and continues to make advances in discovering more about the human body. Because the field became so popular, many institutions and labs have opened over the last century and people continue doing research. With the Creation of the American Society of Bio-mechanics in 1977, the field continues to grow and make many new discoveries.

In the 19th century Étienne-Jules Marey used cinematography to scientifically investigate locomotion. He opened the field of modern 'motion analysis' by being the first to correlate ground reaction forces with movement. In Germany, the brothers Ernst Heinrich Weber and Wilhelm Eduard Weber hypothesized a great deal about human gait, but it was Christian Wilhelm Braune, who significantly advanced the science using recent advances in engineering mechanics. During the same period, the engineering mechanics of materials began to flourish in France and Germany under the demands of the industrial revolution. This led to the rebirth of bone biomechanics when the railroad engineer Karl Culmann and the anatomist Hermann von Meyer compared the stress patterns in a human femur with those in a similarly shaped crane. Inspired by this finding Julius Wolff proposed the famous Wolff's law of bone remodeling.

1.3 Goals of Exercise and Sport Biomechanics

1.3.1 Goals

The goals of exercise and sport biomechanics are performance improvement, as well as injury prevention and rehabilitation. Methods to improve performance include improvement of technique, as well as analysing the equipment used. Biomechanics can lead to modifications in training which could lead to performance benefits. Analysis of technique could isolate deficiencies whereby one could identify the type of conditioning (e.g., strength requirements) the individual need. Biomechanics is useful to assist sports medicine professionals in identifying what forces could cause an injury, how injuries could be prevented, and what exercises may assist with rehabilitation from an injury. Work from the Exercise and Sport Biomechanics Laboratory currently focus on analysis of sport skills, running mechanics, analysis of functional movements for application in Biokinetics, as well as tagging and game analysis for application in team sports.

- In a wider context the goal of biomechanics of sport and physical exercise is also to increase physical fitness. For instance the correct biomechanics of running allows athletes to carry out regular physical exercise for long enough periods of time without being seriously limited by injuries and their consequences.
- The secondary goal of sport biomechanics is strongly related to the main goal because a healthy athlete will perform better than an athlete plagued with frequent injuries.

Technique Improvement

The most common method for improving performance in many sports is to improve an athlete's technique. This is highlighted here as one motivation for studying biomechanics, and it is probably what you thought of when asked how a biomechanist goes about trying to improve an athlete's performance.

The application of biomechanics to improve technique may occur in two ways: Teachers and coaches may use their knowledge of mechanics to correct actions of a student or athlete in order to improve the execution of a skill, or a biomechanics researcher may discover a new and more effective technique for performing a sport skill. In the first instance, teachers and coaches use qualitative biomechanical analysis methods in their everyday teaching and coaching to effect changes in technique. In the second instance, a biomechanics researcher uses quantitative biomechanical analysis methods to discover new techniques, which then must be communicated to the teachers and coaches who will implement them.

Let's look at a simple example of the first case. As a coach, suppose you observe that your gymnast is having difficulty completing a double somersault in the floor exercise. You might suggest three things to the gymnast to help her successfully complete the stunt: (1) jump higher, (2) tuck tighter, and (3) swing her arms more vigorously before takeoff. These suggestions may all result in improved performance and are based on biomechanical principles. Jumping higher will give the gymnast more time in the air to complete the somersault.

Fig arms used in swimming

Tucking tighter will cause the gymnast to rotate faster due to conservation of angular momentum. Swinging the arms more vigorously before takeoff will generate more angular momentum, thus also causing the gymnast to rotate faster. In general, this is the most common type of situation in which biomechanics has an effect on the outcome of a skill. Coaches and teachers use biomechanics to determine what actions may improve performance.

Fig rotating arm fast by gymnast

The second general situation in which biomechanics contributes to improved performance through improved technique occurs when biomechanics researchers develop new and more effective techniques.

Fig Jumping higher

Despite the common belief that new and revolutionary techniques are regularly developed by biomechanists, such developments are rare. Perhaps the reason is that biomechanics as a discipline is a relatively new science. The much more common outcome of biomechanics research is the discovery of small refinements in technique. One example of biomechanics research that did greatly affect the technique and performances in a sport occurred in swimming in the late '60s and early '70s. Research done by Ronald Brown and James "Doc" Counsilman (1971) indicated that the lift forces acting on the hand as it moved through the water were much more important in propelling a swimmer through the water than previously thought. This research indicated that rather than pulling the hand in a straight line backward through the water to produce a propulsive drag force, the swimmer should move the hand back and forth in a sweeping action as it is pulled backward to produce propulsive lift forces as well as propulsive drag forces. This technique is now taught by swimming teachers and coaches throughout the world.

Other examples of sports in which dramatic changes in technique produced dramatic improvement in performance include javelin throwing, high jumping, and cross-country skiing. In 1956, before the Summer Olympic Games in Melbourne, Felix Erasquin, a 48-year-old retired discus thrower from the Basque region of Spain, experimented with an unconventional way of throwing the javelin. Erasquin had experience in barra vasca, a traditional Basque sport that involved throwing an iron bar called a palanka. A turn was used to throw the palanka, and Erasquin incorporated this turn in his innovative javelin throwing technique. Rather than throwing using the conventional technique—over the shoulder with one hand from a run—Erasquin held the javelin with his right hand just behind the grip. The tip of the javelin pointed down to his right, and the tail was behind his back and pointed upward. During the run-up, Erasquin spun around like a discus thrower and slung the javelin from his right hand, which guided the implement. To reduce the frictional forces acting on the javelin as it slid through his hand, it had been dunked in soapy water to make it slippery. The outstanding results achieved by Erasquin and others with this technique attracted international attention. Several throwers using this "revolutionary" technique recorded throws that were more than 10 m beyond the existing javelin world record. Officials at the International Amateur Athletic Federation (IAAF), the governing body for track and field, became so alarmed that they altered the rules for the event, and this unconventional technique became illegal. None of the records set with the Spanish technique were recognized as official world records.

In the late '70s, Bill Koch, an American cross-country skier, began experimenting with a new skating technique he had observed marathon skiers using in Europe. The technique he experimented with was much different from the traditional diagonal stride skiing technique in which cross-country skiers moved their skis parallel to each other in set tracks. In the 1976 Olympic Games in Innsbruck, Austria, Koch surprised the world by winning a silver medal in the 30K cross-country skiing event. More surprising were his performances in the 1982 to 1983 season, when he became the first American ever to win the World Cup. Koch used the skating technique in achieving this title. By the mid-1980s, the skating technique was used by virtually by all elite Nordic ski racers. Beginning with the 1992 Winter Olympics, there were separate competitions for traditional (diagonal stride) and freestyle (skating) cross-country skiing.

With the exception of the swimming example, these examples of new and dramatically different techniques leading to improved performances all happened without the apparent assistance of biomechanics. Maybe this is evidence of the skill of teachers, coaches, and athletes. Through repeated observation, trial and error, and possibly some application of mechanical principles, they have successfully developed excellent techniques for performing skills in most sports without the assistance of biomechanical researchers. But perhaps these improved techniques would have been developed sooner if more teachers and coaches had a working knowledge of biomechanics.

1.3.2 Biomechanics in Sport

Biomechanics in Sport incorporates detailed analysis of sport movements in order to minimise the risk of injury and improve sports performance. Sport and Exercise Biomechanics encompasses the area of science concerned with the analysis of the mechanics of human movement. It refers to the description, detailed analysis and assessment of human movement during sport activities.

Mechanics is a branch of physics that is concerned with the description of motion/movement and how forces create motion/movement. In other words sport biomechanics is the science of explaining how and why the human body moves in the way that it does. In sport and exercise that definition is often extended to also consider the interaction between the performer and their equipment and environment. Biomechanics is traditionally divided into the areas of kinematics which is a branch of mechanics that deals with the geometry of the motion of objects, including displacement, velocity, and acceleration, without taking into account the forces that produce the motion while kinetics is the study of the relationships between the force system acting on a body and the changes it produces in body motion. In terms of this there are skeletal, muscular and neurological considerations we also need to consider when describing biomechanics.

1.3.3 Application

According to Knudson, human movement performance can be enhanced in many ways as effective movement encompasses anatomical factors, neuromuscular skills, physiological capacities and psychological/cognitive abilities. Biomechanics is essentially the science of movement technique and as such tends to be most utilised in sports where technique is a dominant factor rather than physical structure or physiological capacities. The following are of the areas where Biomechanics is applied to either support performance of athletes or solve issues in sport or exercise:

- The identification of optimal technique for enhancing sports performance
- The analysis of body loading to determine the safest method for performing a particular sport or exercise task
- The assessment of muscular recruitment and loading
- The analysis of sport and exercise equipment e.g., shoes, surfaces and racquets.
- Biomechanics is utilised to attempt to either enhance performance or reduce the injury risk in the sport and exercise tasks examined.

1.3.4 Principles

It is important to know several biomechanical terms and principles when examining the role of biomechanics in sport and exercise.

Forces and Torques

A force is simply a push or pull and it changes the motion of a body segment or the racket. Motion is created and modified by the actions of forces (mostly muscle forces, but also by external forces from the environment). When Force rotates a body segment or the racket, this effect is called a torque or moment of force.

Example - Muscles create a torque to rotate the body segments in all tennis strokes. In the service action internal roation of the upper arm, so important to the power of the serve, is the result of an internal rotation torque at the shoulder joint caused by muscle actions (latissimus dorsi and parts of the pectoralis major and deltoid). To rotate a segment with more power a player would generally apply more muscle force.

Newton's Laws of Motion

Newton's Three Laws of Motion explain how forces create motion in sport. These laws are usually referred to as the Laws of Inertia, Acceleration, and Reaction.

- Law of Inertia - Newton's First Law of inertia states that objects tend to resist changes in their state of motion. An object in motion will tend to stay in motion and an object at rest will tend to stay at rest unless acted upon by a force. Example - The body of a player quickly sprinting down the field will tend to want to retain that motion unless muscular forces can overcome this inertia.
- Law of Acceleration - Newton's Second Law precisely explains how much motion a force creates. The acceleration (tendency of an object to change speed or direction) an object experiences is proportional to the size of the

force and inversely proportional to the object's mass (F = ma). Example - If a player improves leg strength through training while maintaining the same body mass, they will have an increased ability to accelerate the body using the legs, resulting in better agility and speed. This also relates to the ability to rotate segments, as mentioned above.

- Law of Reaction - The Third Law states that for every action (force) there is an equal and opposite reaction force. This means that forces do not act alone, but occur in equal and opposite pairs between interacting bodies. Example - The force created by the legs "pushing" against the ground results in ground reaction forces in which the ground "pushes back" and allows the player to move across the court (As the Earth is much more massive than the player, the player accelerates and moves rapidly, while the Earth does not really accelerate or move at all). This action-reaction also occurs at impact with the ball as the force applied to the ball is matched with an equal and opposite force applied to the racket/body.

Momentum

Newton' Second Law is also related to the variable momentum, which is the product of an object's velocity and mass. Momentum is essentially the quantity of motion an object possesses. Momentum can be transferred from one object to another. There are different types of momentum which each have a different impact on the sport.

- Linear Momentum, which is momentum in a straight line. Example - Linear momentum is created as the athlete sprints in a straight line down the 100m straight on the track.
- Angular Momentum, which is rotational momentum and is created by the rotations of the various body segments e.g. The open stance forehand uses significant angular momentum. The tremendous increase in the use of angular momentum in ground strokes and serves has had a significant impact on the game of tennis. One of the main reasons for the increase in power of the game today is the incorporation of angular momentum into groundstroke and serve techniques. In tennis, the angular momentum developed by the coordinated action of body segments transfers to the linear momentum of the racket at impact

Center of Gravity

The Center of Gravity (COG) is an imaginary point around which body weight is evenly distributed The center of gravity of the human body can change considerably because the segments of the body can move their masses with joint rotations. This concept is critical to understanding balance and stability and how gravity affects sport techniques.

The direction of the force of gravity through the body is downward, towards the centre of the earth and through the COG. This line of gravity is important to understand and visualise when determining a person's ability to successfully maintain Balance. When the line of gravity falls outside the Base of Support (BOS), then a reaction is needed in order to stay balanced.

The center of gravity of a squash racket is a far simpler process and can usually be found by identifying the point where the racket balances on your finger or another narrow object.

Balance

Balance is the ability of a player to control their equilibrium or stability. You need to have a good understanding of both static and dynamic balance:

- Static Balance - The ability to control the body while the body is stationary. It is the ability to maintain the body in some fixed posture. Static balance is the ability to maintain postural stability and orientation with centre of mass over the base of support and body at rest.
- Dynamic Balance - The ability to control the body during motion. Defining dynamic postural stability is more challenging, Dynamic balance is the ability to transfer the vertical projection of the centre of gravity around the supporting base of support. Dynamic balance is the ability to maintain postural stability and orientation with centre of mass over the base of support while the body parts are in motion.

1.3.4 Lower Limb Biomechanics

As humans, ambulation is our main form of movement, that is, we walk upright and are very reliant on our legs to move us about. How the foot strikes the ground and the knock on effect this has up the lower limbs to the knee, hips, pelvis and low back in particular has become a subject of much debate and controversy in recent years.

Lower limb biomechanics refers to a complex interplay between the joints, muscles and nervous system which results in a certain patterning of movement, often referred to as 'alignment'. Much of the debate centres around what is considered 'normal' and what is considered 'abnormal' in biomechanical terms as well as the extent to which we should intervene should abnormal findings be found on assessment. This section examines the biomechanics of the lower extremity in particular the anatomy and biomechanics of the foot and ankle, the impact of Q Angle on tech mechanics of the Hip and Knee and finally the implications of this on gait.

Foot & Ankle Biomechanics

The foot and ankle form a complex system which consists of 26 bones, 33 joints and more than 100 muscles, tendons and ligaments. It functions as a rigid structure for weight bearing and it can also function as a flexible structure to conform to uneven terrain. The foot and ankle provide various important functions which includes: supporting body weight, providing balance, shock absorption, transferring ground reaction forces, compensating for proximal malalignment, and substituting hand function in individuals with upper extremity amputation/paralysis all which are key when involved with any exercise or sport involving the lower limbs.

Q Angle

An understanding of the normal anatomical and biomechanical features of the patellofemoral joint is essential to any evaluation of knee function. The Q angle formed by the vector for the combined pull of the quadriceps femoris muscle and the patellar tendon, is important because of the lateral pull it exerts on the patella.

The direction and magnitude of force produced by the quadriceps muscle has great influence on patellofemoral joint biomechanics. The line of force exerted by the quadriceps is lateral to the joint line mainly due to large cross-sectional area and force potential of the vastus lateralis. Since there exists an association between patellofemoral pathology and excessive lateral tracking of the patella, assessing the overall lateral line of pull of the quadriceps relative to the patella is a meaning.

1.3.5 Upper Limb Biomechanics

Correct biomechanics are as important in upper limb activities as they are in lower limb activities. The capabilities of the upper extremity are varied and impressive. With the same basic anatomical structure of the arm, forearm, hand, and fingers, major league Baseball Pitchers pitch fastballs at 40 m/s, Swimmers cross the English Channel, Gymnasts perform the iron cross, and Olympic Boxers in weight classes ranging from flyweight to super heavyweight showed a range of 447 to 1,066 pounds of peak punching force.

The structure of the upper extremity is composed of the shoulder girdle and the upper limb. The shoulder girdle consists of the scapula and clavicle, and the upper limb is composed of the arm, forearm, wrist, hand, and fingers. However, a kinematic chain extends from the cervical and upper thoracic spine to the fingertips. Only when certain multiple segments are completely fixed can these parts possibly function independently in mechanical roles.

Scapulohumeral Rhythm

Scapulohumeral rhythm (also referred to as glenohumeral rhythm) is the kinematic interaction between the scapula and the humerus, first published by Codman in the 1930s. This interaction is important for the optimal function of the shoulder. When there is a change ment of the normal position of the scapula (describe) relative to the humerus, can this can cause a disfunction dysfunction of the scapulohumeral rhythm. The change ment of the normal position is also called scapular dyskinesia. Various studies of the mechanism of the shoulder joint have attempted to describe the global motion capacity of the shoulder; refer to that description, can you evaluate the shoulder to see if the function is correct? Also, explain the complex interactions between components involved in placing the hand in space.

1.4 Sport Specific Biomechanics

1.4.1 Running Biomechanics

Running is similar to walking in terms of locomotor activity. However, there are key differences. Having the ability to walk does not mean that the individual has the ability to run. There are some differences between the gait and run cycle - the gait cycle is one third longer in time, the ground reaction force is smaller in the gait cycle (so the load is lower), and the velocity is much higher. In running, there is also just one stance phase while in stepping there are two. Shock absorption is also much larger in comparison to walking. This explains why runners have more overload injuries. Running Requires:

- ❖ Greater Balance
- ❖ Greater Muscle Strength
- ❖ Greater Joint Range of Movement

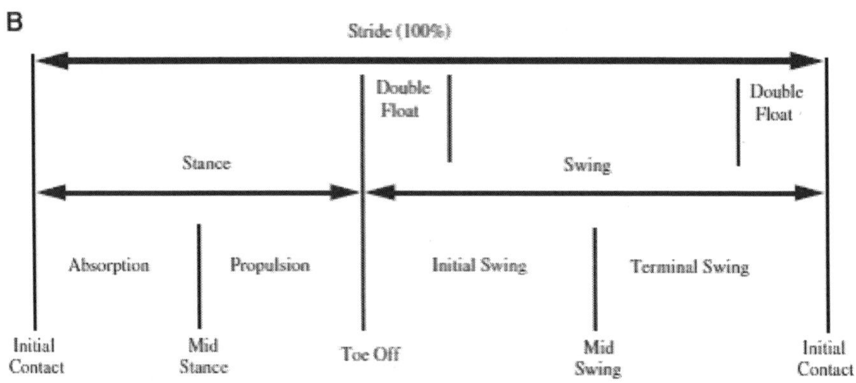

Fig Running biomechanics

There is a need for greater balance because the double support period present in walking is not present when running. There is also the addition of a double float period during which both feet are off the ground, not making contact with the support surface. The amount of time that the runner spends in float increases, as the runner increases in speed. The muscles must produce greater energy to elevate the head, arms and trunk (HAT) higher than in normal walking, and to support HAT during the gait cycle. The muscles and joints, must also be able to absorb increased amounts of energy to control the weight of HAT. During the running gait cycle, the

Ground reaction force (GRF) at the centre of pressure (COP) have has been shown to increase to 250% of the body weight.

Fig Biomechanics of Running

Joint Motion

- Beginning of stance phase - hip is in about 50deg flexion at heel strike, continuing to extend during the rest of the stance phase. I reaches 10deg of hyperextension after toe off.
- The hip flexes to 55deg flexion in the late swing phase.
- Before the end of the swing phase, the hip extends to 50 deg to prepare for the heel strike.
- he knee flexes to about 40deg as the heel strikes, then flexes to 60deg during the loading response
- The knee begins to extend after this, and reaches 40deg flexion just before toe-off.
- During swing phase and the initial part of the float period, the knee flexes to reach maximum flexion of 125deg during the mid-swing.
- The knee then prepares for heel strike by extending to 40deg
- The ankle is in about 10deg of dorsiflexion when the heel strikes, and then dorsiflexes rapidly to 25deg DF.
- Plantarflexion happens almost immediately, continuing throughout the rest of the stance phase of running, and as it enters swing phase also.
- Plantar flexion reaches a maximum of 25deg in the first few seconds of ~~of~~ swing phase.
- The ankle then dorsiflexes throughout the swing phase to 10deg in the late stage of swing phase, preparing for heel strike.

- The lower limb medially rotates during the swing phase, continuing to medially rotate at heel strike.
- The foot pronates at heelstrike.
- Lateral rotation of the lower limb stance leg begins as the swing leg passes by the stance leg in mid stance position.

Muscle Activity

- Gluteus maximus and gluteus medius are both active at the beginning of stance phase, and also at the end of swing phase.
- TFL is active from the beginning of stance, and also the end of swing phase. It is also active between early and mid-swing.
- Adductor Magnus is active for about 25% of cycle, from late stance to early part of swing phase.
- Iliopsoas activity occurs during swing phase for 35-60% of cycle.
- Quadriceps works in an eccentric manner for the initial 10% of the stance phase. Its role is to control knee flexion as the knee goes through rapid flexion. It stops being active after the first part of the stance phase, there is then no activity until the last 20% of swing phase. At this point it becomes concentric in behaviour so it can extend the knee to prepare for heel strike.
- Medial Hamstrings become active at the beginning of the stance phase (18-28% of stance), they are also active throughout much of the swing phase (40-58% of initial swing then the last 20% of swing). They acting to extend the hip and control the knee through concentric contraction in late swing, the hamstrings act eccentrically to control knee extension and take the hip into extension again.
- Gastrocnemius muscle activity starts just after loading at heel strike, remaining active up until 15% of the gait cycle (this is where its activity begins in walking). It then re-starts its activity in the last 15% of the swing phase.
- Tibialis anterior muscle is active through both stance and swing phases in running. It is active for about 73% of the cycle (compared to 54% when walking). The swing phase when running, is 62% of the total gait cycle, compared to 40% when walking, so TA is considerably more active when running. Its activity is mainly concentric or isometric, enabling the foot to clear the support surface during the swing phase of the running gait.

1.4.2 Cycling Biomechanics

Cycling was initially invented by Baron Carl von Drais in 1817, but not as we know it. This was a machine which intailly initially had two wheels that were connected by a wooden plank with a rudder device for steering. It involved people running along the ground whilst sitting down; giving them the name of a 'running machine' (in all senses) or a velocipied velocipede. This was solely used by the male population at the time of invention. The velocipied velocipede then made a huge design development in the 1860's at the Michaux factory in Paris. They added leaver arms to the front wheel which were prepelled propelled by pedals at the feet. This was the first conventional bicycle, and since then and up until the current day the bicycle has made great design and technological advances.

A survey in 2014 estimated that over 43% have or have an access to a bike and 8% of the population aged 5 and above cycled 3 or more times a week. With such a large amount of people cycling, whether it be professional, recreational or for commuting this increase the chance of developing an injury, so it is time we understood the biomechanics of cycling.

Points of Contact

There are 3 points of contact in cycling. Meaning 3 points of the body that make contact with the bike:

- Pelvis on the saddle
- Hand on the handlebars
- Foot on the pedal

Something to be aware of is that these areas can undergo sustained amounts of pressure and compression which can cause numbness, pain and weakness.

Phases of Cycling / Pedaling

There are 2 main phases of the pedal cycle; the power phase and the recovery phase. If you imagine the pedal cycle as a clock face and you start with the pedal at 12 o'clock, this is known as Top Dead Centre (TDC). The pedal is then pushed down from 12 until 6 o'clock, this position is known as Bottom Dead Centre (BDC). The phase between the 2 is known as the Power Phase where all the force is generally generated to propel the bike forward.

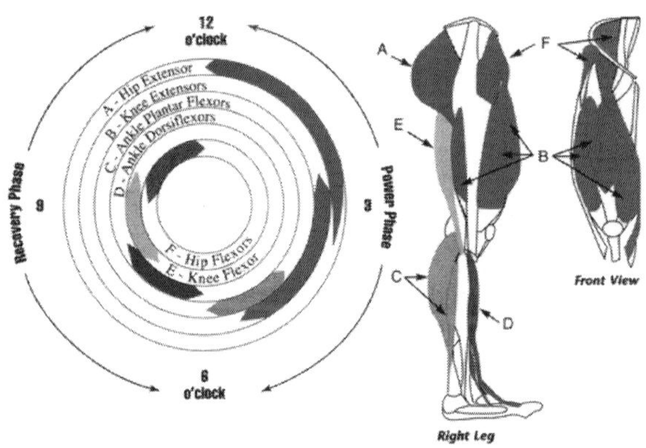

Fig Anatomy of movements

The transition from BDC back up to TDC is known as the Recovery Phase. Now not all of the muscles just switch of during this phase, it just is not as active as the Power Phase. If you look at the image you can see what muscles are working at which points during the pedal cycle.

Anatomy of Cycling

There are many components that are working whilst cycling, this is not just a sport of the lower limbs. Below the areas of anatomy have been broken down to provide more detail on what is happening and when.

Fig Working of limbs

The pelvis is the start of the lower limb complex, and is compromised of the ischium and the ilium. The ischial tuberosities (also referred to at the sitting bones) are located here and play an important roll role for the hamstrings, as this is where all three origination. The hip is also an important anatomical feature as this is a large 'ball and socket' type joint, which allows for a large degree of multidirectional movement.

During cycling the hip allows for and guides hip flexion, extension and small degree of rotation. Further down the lower limb complex the knee is found. This 'hinge' joint acts as a lever to the femur, as ~~as~~ the femur is the longest bone in the body this can create large amounts of torque. This is where the patella plays a vital role, as it acts as a fulcrum and enables the force from the upper leg to be transferred to the lower leg. The patella is a sesamoid bone that sits within the patella tendon and connects the quadriceps to the tibial tuberosity. The patella glides in the intercondylar fosse of the femur.

Moving down the lower limb complex the next main joint of relevance is the ankle. This joint allows for dorsi-flexion and plantar-flexion in cycling, which allows for a term known as 'ankleing' where the foot moves from a dorsi-flexed position to a plantar-flexed position through the bottom of the pedal stroke before returning back to a dorsi-flexed position.

The foot has many small joints but primarily this is where the force that is generated from the lower limb complex is transferred to the pedal. Irregular amounts of force or compression running through the foot can result in neural pain and tissue damage from compression.

The initiation of the pedal cycle starts with the gluteals, taking the hip from a flexed position at TDC through the power phase to an extended position. Then at approximately 3 o'clock in the pedal cycle the quadriceps kick in to take the knee from it's its flexed position to an extended position at BDC.

The rectus femoris is one of the four quadricep muscles, but the only one to cross both the hip and knee joint, giving it duel responsibility of hip flexion and knee extension. The quadriceps work in close partnership with the gluteals (maximus), these are two large powerful groups of muscles produce the greatest amount of torque in cycling. Due to the position on the bike, muscles such as the rectus femoris can become shortened leading to anterior hip pain, but also commonly patella femoral pain. This is due to the rectus femoris leading into the patella tendon and attaching onto the tibial tiberosity, so if this muscle becomes shortened it can increase the compressive forces around the patella, causing discomfort.

The main role of the hamstrings is knee flexion but they also assist hip extension. During cycling, depending on the position that is adopted by the cyclist if on an upright bike, the ischial tuberosities may take most of the load through the saddle therefore compressing the origin of the hamstrings. The hamstrings main role in cycling is to assist the knee flexion up through the back part of the pedal stroke but they also play an important part in stabilising the knee through BDC.

Further down the chain into the lower leg the gastrocnemius and soleus haven't been found to add much power to the pedal stroke but their main role is to stabilise the lower leg to enable an efficient transition of the force generated by the upper leg to the pedal.

Then as we come back up the other side of the pedal cycle the hamstrings kick in to help flex the knee to bring it back up to TDC, and when enough knee flexion and hip flexion is achieved the hip flexors start to work again to bring the leg up to true TDC to start the cycle again.

There is more to cycling than the movement of the lower limbs. The trunk and back play an important role in stabilising the spine and maintaining posture. There are many muscles within the back but to name a few multifidi and the quadratus lumborum are a couple of the main stabilisers of the spine when undergoing lateral and rotational movements, such as right leg enters the power phase whilst the left side of the spine stabilises and vice versa. The erector spinae muscles also play an important part in maintaining a stable posture whilst on the bike.

Abdominal muscles such as the rectus abdominus help to maintain stability as does the obliques. The obliques similarly to the back muscles will help stabilise a contralateral limb movement. As we move up the spine toward the shoulders, the latissimus dorsi and trapezius muscles enable the rider to fix their upper body onto the handlebars. The upper body has a role in stabilising contralateral torque, so as the right leg pushes down the left arm anchors to the handle bars and pulls up.

1.4.3 Baseball Pitching Biomechanics

Baseball pitching is one of the most intensely studying athletic motion. Although focus has been more on shoulder, entire body movement is required to perform a baseball pitching. Throwing is also considered one of the fastest human motions performed, and maximum humeral internal rotation velocity reaches about 7000 to 7500/second.

1.4.4 Tennis Biomechanics

Tennis biomechanics is a very complex task. Consider hitting a tennis ball. First the athlete needs to see the ball coming off their opponent's racquet. Then, in order, they have to judge the speed, spin, trajectory and, most importantly, the direction of the tennis ball. The player then needs to adjust their body position quickly to move around the ball. As the player returns prepares to hit the ball the body is in motion, the ball is moving both in a linear and rotation direction if there is pin on the ball, and the racquet is also in motion. The player must coordinate all these movements in approximately a half a second so they strike the ball as close to the centre of the racquet in order to produce the desired spin, speed and direction for return of the ball. A mistake in any of these movements can create an error.

Chapter 2

External Biomechanics: Forces in Maintaining Equilibrium or Changing Motion

2.1 Forces

Force is simply a push or a pull exerted by one object on another. For an example, when a muscle contracts it creates a force that is transferred to the tendon to pull on the bony attachment, thus resulting in motion. An important consideration with force is that it is a vector quantity.

A vector quantity always has 3 variables:

- ❖ A Point of Application
- ❖ A Magnitude
- ❖ A Direction

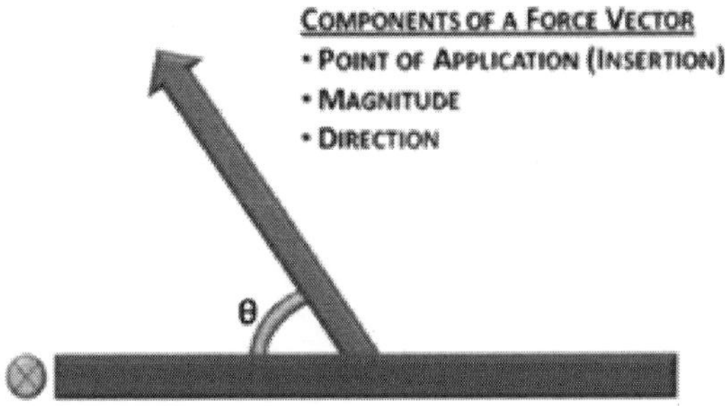

Fig Simple Force Diagram

Force is an elementary concept in biomechanics, therefore it will be explained in an individual chapter. Without applying force we would be completely helpless in sport and physical exercise. Forces make us move, stop, or change direction. Forces are even important also when we stand still. For example we exert forces to keep our upright posture. To keep balance while riding a bicycle or skiing also requires to be able to control various forces.

If we omit inertia forces, force is a quantitative description of the interaction between two physical bodies, such as an object and its environment. For the needs of biomechanics we can state, in a simplified way, that force is produced by pull, push, or gravity. Physical bodies in contact always exert forces at each other. Forces always occur in pairs. Every action is accompanied by a reaction of equal magnitude but opposite direction.

2.1.1 Characteristics of Force

The four features of a force are its magnitude, action line, direction, and point of application. A force cannot be described unless these factors are known because any variation of one or more of these factors changes the result.

Magnitude And Action Line

Magnitude is a scalar quantity such as time, speed, temperature, volume, and length that has no direction. More factors must be known besides magnitude if a force involving a stress is to be accurately described. For example, a 5-lb weight held in a hand when the arm is held vertically produces a far different effect in the shoulder joint than the same weight held in the hand when the arm is held horizontally. Thus, the action line (line of force application) must be known.

Direction And Point Of Application

Since a pulling force has a different effect than a pushing force, it is important to know the direction of force along the action line.

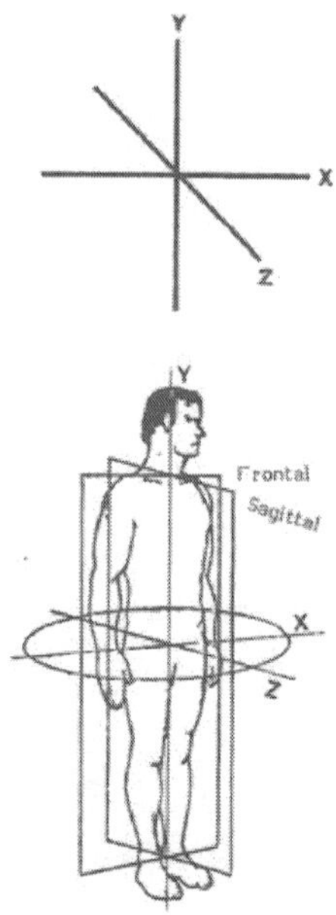

Fig Direction of forces

Vectors

Although force is usually applied over an area, it is usually described in biomechanical drawings as a summarized point force by an arrow. Any quantity that gives both magnitude and direction is a vector (e.g. a force) that can be described by a straight line. Quantities that involve only magnitude are referred to as scalars. When illustrating a force, the vector's length should be proportional to the magnitude of the force. For example, if 1 inch is used to represent a 10-lb force, a 2-inch line would represent a 20-lb force.

A vector can be used to define a force in a simple line drawing if the vector drawn to scale represents magnitude by the line's length, if the vector's tail indicates the point where the force is applied on the object, and if the direction of force is indicated by the vector's arrowhead. If the magnitude of a vector is known, it should be indicated (eg, 1 inch = 20 lb). If the magnitude is not known, it is indicated by the capital letter F (force) or P (pressure) to designate the unknown magnitude. Distances are usually represented by lower case letters.

The force of gravity is always directed toward the center of the earth. Thus, gravity's line of action and direction are constants. In the upright "rigid" body, the gravitational force on the entire mass can be thought of as a single vector through the center of mass which represents the sum of many parallel positive and negative coordinates (Fig. 2.4). If a weight is held in the outstretched hand, the quantity of gravitational force is governed by the weight of the extremity plus the weight held.

Space

As a force may act along a single line in a single plane or in any direction in space, this must be considered to provide an illustrative reference system. In a two-dimensional system, the plane is simply divided into four quadrants by means of a perpendicular vertical ordinate line (Y axis) and a horizontal abscissa line (X axis). The point of axial intersection is referred to as the system's origin. Abscissa (X) measurements to the right of the origin are considered positive, while those to the left are negative. Ordinate (Y) measurements above the origin are considered positive, while those below the origin are negative. By this method, any point on the plane can be given an X and Y value.

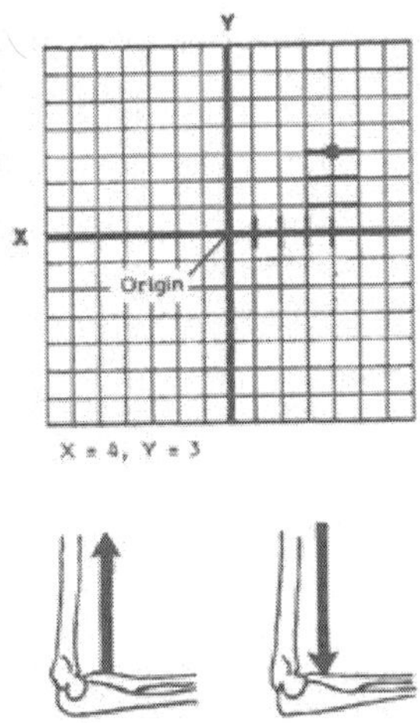

Fig Vector position in space

The term coordinates refers to specific point locations from the origin which have been given a value. For example, a point located 5 units to the right of origin and 3 units down from the origin would be defined as X = 5, Y = 3.

A third axis (usually titled Z) can be introduced to locate points in three dimensions. Such an axis crosses the origin and is perpendicular to the other two planes (X and Y). All Z points in front of the X-Y plane are positive, while those behind are negative. By utilizing X, Y, and Z coordinates, any point in space can be located and depicted. However, a minimum of six coordinates is necessary to specify the position of a rigid body. Force and moment are three dimensional vectors having three components each; thus load may be considered a six-component vector.

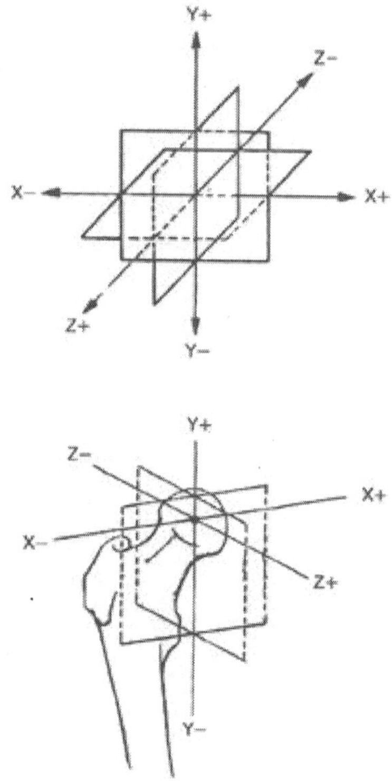

Fig System of coordinates

In biomechanics, the body's origin is located at the body's center of mass which is usually just anterior to the S2 segment. When this point is known, gross body space can be visualized as being in the sagittal (right-left) Y-Z plane, frontal or coronal (anterior-posterior) X-Y plane, or horizontal or transverse (superior-inferior) X-Z plane. With such a reference system, any movement of any body segment in these planes can be approximately described by placing a coordinate system at the axis of a joint and projecting the action lines of the muscles involved.

2.2 Classifying Forces

Force is the term used to describe interactions between objects causing changes in velocity. Forces are either repulsive (pushing) or attractive (pulling). There are four fundamental forces: gravitational, electromagnetic, and strong nuclear, and weak nuclear.

Other forces that can be observed are functions of these fundamental forces. In biomechanics, these include the normal force, friction, and elasticity.

Newton's three laws describe motion as follows: (1) objects remain at rest or continue moving at constant velocity unless a force acts upon them, (2) the rate of change of velocity of an object as a result of an applied force is proportional to its mass, and (3) when an object applies a force to a second object, the second object applies an equal and opposite force to the first.

Ground reaction forces are normal forces that are produced when objects come into contact with the ground. They are exerted by the ground in response to the action forces exerted by the object.

Force plates or force platforms measure the ground reaction forces in response to the action forces exerted by the feet of individuals who stand on them, jump onto them, or step on them while walking, running or sprinting. The forces are recorded as components in the vertical, horizontal and medio-lateral directions.

Force treadmills are treadmills with integrated force plates or force platforms that measure the ground reaction forces in response to the action forces exerted by the feet of individuals who step on them while walking, running or sprinting.

Despite concerns, force treadmills can be used to measure horizontal ground reaction forces reliably, as the internal forces within the treadmill cancel each other out, so long as the whole treadmill is mounted upon the force transducers (as in most common models).

Treadmill running involves a lower oxygen cost than overground running because there is no wind resistance. Using a gradient of 1% equates the oxygen costs of running on a treadmill and running overground.

When a familiarisation period of >6 minutes is used, joint angle movements are very similar between running at a constant speed on a treadmill and running overground, as theoretical analysis using a moving frame of reference would suggest.

Fundamental Forces

Force is the term used to describe interactions between objects that cause changes in velocity. Forces can be either repulsive (pushing) or attractive (pulling). Researchers working in the field of physics have identified that there are four fundamental forces: gravitational, electromagnetic, strong nuclear, and weak nuclear. Gravitational force (also known as gravity) is an attractive force between objects that have mass. Electromagnetic force can be either attractive or repulsive (depending on the charge) and acts between electrically charged ions. The strong and weak nuclear forces describe the interactions between subatomic particles.

Derived Forces

Other forces that can be observed in nature are simply functions of these fundamental forces and these can be described as derived forces. In biomechanics, there are many important forces that are derived from these fundamental forces, including the normal force, friction, and elasticity. The normal force is extremely important in biomechanics, as it describes the interaction between objects that come into contact with one another. In this context, "normal" refers to the direction of force, which is directly opposing the direction of approach of the two objects. Normal forces act when atomic particles come within range of one another and are repulsive. They therefore act in order to repel other objects directly away from them. However, they are only active within a very short range. Friction acts at right angles (perpendicular) to the normal force and describes the interaction between objects that move adjacent to one another. Again, this force opposes the prevailing direction of motion. Elasticity in biomechanics often describes a simple phenomenon that is made up of a complex set of interactions. Elastic tissues, such as muscles and tendons, display a tendency to return to their starting length after being elongated, which is the property of elasticity. However, there are many different strings of molecules involved in this process, all of which contribute to the overall elastic property of the tissue, and different materials have different bases for their elastic properties.

There many types of forces to consider when looking at biomechanics. A detailed consideration of these forces is necessary to achieve a full picture of the biomechanical system. There are 4 types of forces to consider with biomechanics mentioned below. However, this is by no means an exhaustive list of all the forces involved with biomechanics.

2.2.1 Motion Forces -Rotatory, Translatory, Curvilinear

Rotary motion:

It is an angular motion of a body can take place about an axis that passes through the body or about an axis that does not pass through the body. A gymnast swinging on the rings is an example of rotary motion about an axis that does not pass through the body. Figure skater rotating on ice rink is an example of rotary motion about an axis that passes through the body.

Curvilinear Motion

Curvilinear motion is defined as motion that occurs when a particle travels along a curved path. The curved path can be in two dimensions (in a plane), or in three dimensions. This type of motion is more complex than **rectilinear (straight-line) motion**. Three-dimensional curvilinear motion describes the most general case of motion for a particle.

To find the velocity and acceleration of a particle experiencing curvilinear motion one only needs to know the position of the particle as a function of time. Let's say we are given the position of a particle *P* in three-dimensional Cartesian (*x,y,z*) coordinates, with respect to time, where

$$x_p = x_p(t)$$
$$y_p = y_p(t)$$
$$z_p = z_p(t)$$

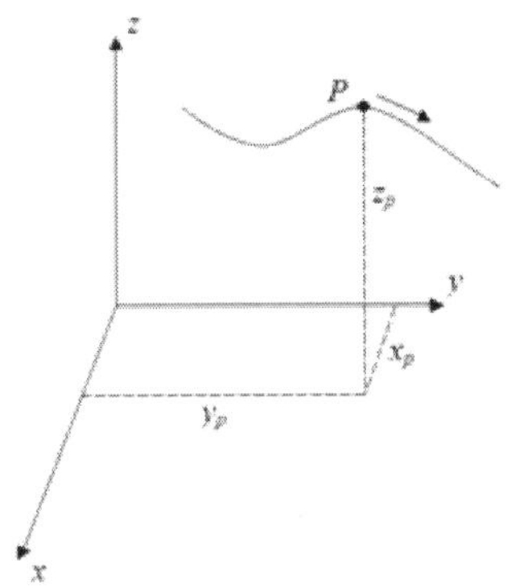

The velocity of the particle *P* is given by

$$v_x = \frac{dx_p}{dt}$$
$$v_y = \frac{dy_p}{dt}$$
$$v_z = \frac{dz_p}{dt}$$

The acceleration of the particle *P* is given by

$$a_x = \frac{d^2x_p}{dt^2}$$
$$a_y = \frac{d^2y_p}{dt^2}$$
$$a_z = \frac{d^2z_p}{dt^2}$$

As you can see, if we know the position of a particle as a function of time, it is a fairly simple exercise to find the velocity and acceleration. You simply take the first derivative to find the velocity and the second derivative to find the acceleration. The magnitude of the velocity of particle *P* is given by

$$v_p = \sqrt{v_x^2 + v_y^2 + v_z^2}$$

The magnitude of the acceleration of particle *P* is given by

$$a_p = \sqrt{a_x^2 + a_y^2 + a_z^2}$$

Note that the direction of velocity of the particle *P* is always tangent to the curve (i.e. the path traveled, denoted by the curve in the figure). But the direction of acceleration is generally *not* tangent to the curve.

However, the acceleration component tangent to the curve is equal to the time derivative of the magnitude of velocity of the particle *P* (along the curve). In other words, if v_t is the magnitude of the particle velocity (tangent to the curve), the acceleration component of the particle tangent to the curve (a_t) is simply

$$a_t = \frac{dv_t}{dt}$$

In addition, the acceleration component normal to the curve (a_n) is given by where *R* is the radius of curvature of the curve at a given point on the curve (x_p,y_p,z_p).

The figure below illustrates the acceleration components a_t and a_n at a given point on the curve (x_p,y_p,z_p).

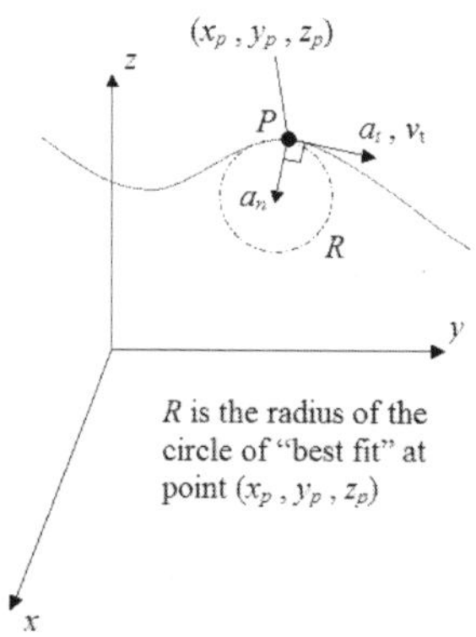

For the specific case where the path of the curve is given by *y* = *f(x)*(two-dimensional motion), the radius of curvature *R* is given by

$$R = \left| \frac{\left(1+\left(\frac{dy}{dx}\right)^2\right)^{3/2}}{\frac{d^2y}{dx^2}} \right|$$

Where, $|x|$ means the "absolute value" of *x*. For example, $|-2.5| = 2.5$, and $|3.1| = 3.1$.

However, it is usually not necessary to know the radius of curvature *R* along a curve. But nonetheless, it is informative to understand it on the basis of its relationship to the normal acceleration (a_n).

In Polar Coordinates

It is sometimes convenient to express the planar (two-dimensional) motion of a particle in terms of polar coordinates (R,θ), so that we can explicitly determine the velocity and acceleration of the particle in the radial (*R*-direction) and circumferential (θ-direction). For this type of motion, a particle is only allowed to move along the radial *R*-direction for a given angle θ.

For a particle *P* defined in polar coordinates (as shown below), we can derive a general equation for its radial velocity (v_r), radial acceleration (a_r), circumferential velocity (v_c), and circumferential acceleration (a_c). Note that the circumferential direction is perpendicular to the radial direction.

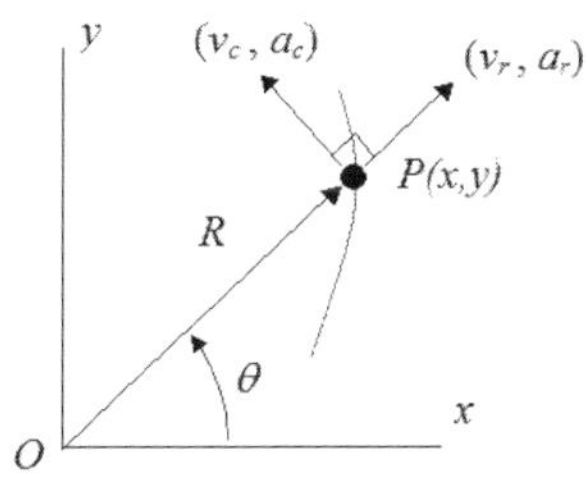

The position of the particle *P* is given with respect to time, where

$$x(t) = R\cos\theta$$
$$y(t) = R\sin\theta$$

To find the velocity, take the first derivative of *x(t)* and *y(t)* with respect to time:

$$\frac{dx}{dt} = \frac{dR}{dt}\cos\theta - R\sin\theta \cdot \frac{d\theta}{dt}$$

$$\frac{dy}{dt} = \frac{dR}{dt}\sin\theta + R\cos\theta \cdot \frac{d\theta}{dt}$$

To find the acceleration, take the second derivative of *x(t)* and *y(t)* with respect to time:

$$\frac{d^2x}{dt^2}=\frac{d^2R}{dt^2}\cos\theta-2\frac{dR}{dt}\sin\theta\cdot\frac{d\theta}{dt}$$
$$-R\cos\theta\cdot\left(\frac{d\theta}{dt}\right)^2-R\sin\theta\cdot\frac{d^2\theta}{dt^2}$$

$$\frac{d^2y}{dt^2}=\frac{d^2R}{dt^2}\sin\theta+2\frac{dR}{dt}\cos\theta\cdot\frac{d\theta}{dt}$$
$$-R\sin\theta\cdot\left(\frac{d\theta}{dt}\right)^2+R\cos\theta\cdot\frac{d^2\theta}{dt^2}$$

Without loss of generality we can evaluate the velocities and accelerations at angle $\theta = 0$, knowing that (at this angle) radial velocity and radial acceleration is in the *x*-direction, and circumferential velocity and circumferential acceleration is in the *y*-direction.

Setting $\theta = 0$ we have:

$$v_r=\frac{dx}{dt}$$

$$v_r=\frac{dR}{dt} \quad (1)$$

$$a_r=\frac{d^2x}{dt^2}$$

$$a_r=\frac{d^2R}{dt^2}-R\left(\frac{d\theta}{dt}\right)^2 \quad (2)$$

$$v_c=\frac{dy}{dt}$$

$$v_c=R\frac{d\theta}{dt} \quad (3)$$

$$a_c=\frac{d^2y}{dt^2}$$

$$a_c=2\frac{dR}{dt}\cdot\frac{d\theta}{dt}+R\frac{d^2\theta}{dt^2} \quad (4)$$

Equations (1), (2), (3), and (4) fully describe the curvilinear motion of a particle *P* in polar coordinates. The term $d\theta/dt$ is called angular velocity. It has units of rad/s. One rad (radian) = 57.296 degrees. The term $d^2\theta/dt^2$ is called angular acceleration. It has units of rad/s^2.

Since v_r and v_c are perpendicular to each other, the magnitude of the velocity of particle *P* is given by

$$v_p = \sqrt{v_r^2 + v_c^2}$$

Since a_r and a_c are perpendicular to each other, the magnitude of the acceleration of particle *P* is given by

$$a_p = \sqrt{a_r^2 + a_c^2}$$

2.2.2 External Forces – Gravity, Wind, Objects, Other People

In our everyday lives, we are exposed to many external physical forces as we perform our activities. These forces vary in magnitude and direction, and hence, vary in their effects. When you feel a nice spring morning breeze, for instance, the feeling differs from the effect of wind loads on high rise buildings.

In structural analysis, external forces are forces applied to the structure from the outside of the structure. Examples of external forces include dead loads, such as the weight of the structure itself and the non-structural materials it supports, and live loads, which include moving loads, such as occupants, goods, and furniture, as well as wind loads, seismic loads, and impact loads, among others.

A reaction force at points where the structural member is supported resulting from external forces is also an external force. In addition, despite the lack of connection between two or more objects, forces such as gravitational and electromagnetic forces can also be classified as external forces.

2.2.3 Internal Forces – Muscles, Connective Tissue (Elastic), Bone

Internal forces are produced from the external forces acting on structure members such as poled, beamed or columned. Generally, we have three types of internal forces: axial, dhear and moment. Axial force, sometimes called 'normal force,' is a compression or tension force acting aligned with the extension of a structure member. Shear force is a force acting in a direction perpendicular to the alignment of the member. Moment force, lastly, is a turning result of a force multiplied by the distance from its acting location to the turning point. The number of these components varies in one-dimensional, two-dimensional and three-dimensional cases. Now the questions are what each of these components do and how to calculate these internal forces.

Equilibrium

When a structure member such as a pole, beam or column is in equilibrium, it means that it isn't moving as it is supposed to be. Therefore, the combination or resultant of all the external forces applied to the member equals zero. In fact, the internal forces maintain the equilibrium of a structure member in different directions.

1D Case

In the simplest example, let's consider a pole in a building structure (1D case). Depending on the applied external loading, this pole can be in either tension or compression. Now, what if the pole is cut from left to right of an arbitrary size (it's called a free body diagram)? Since the pole is not moving, as we discussed, the resultant of all the applied forces, including the external and internal forces, still needs to equal zero.

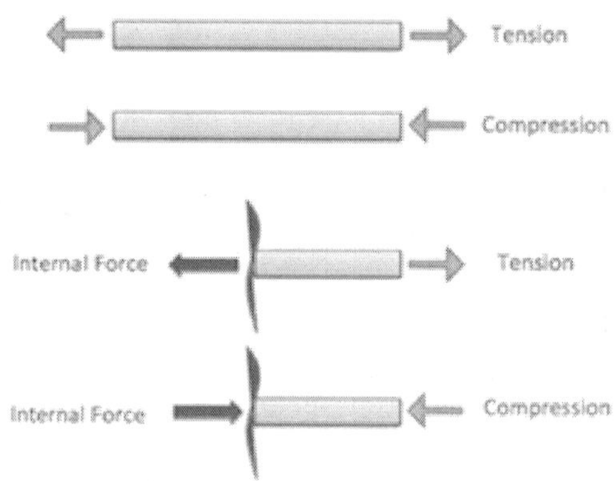

2D Case

We will have the same concept in a 2D case with a beam. Since we assume that the beam is fixed and not moving, the resultant of all applied forces from different directions equal zero.

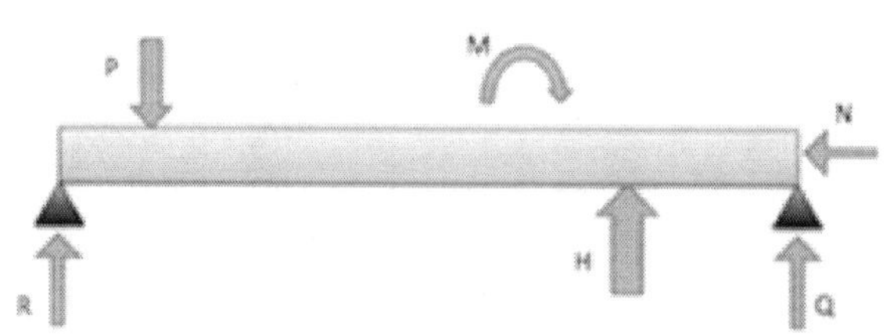

R + P + M + H + Q + N = Zero

3D Case

Similar to the 1D case example, cutting the beam from left or right of an arbitrary size and section will result in a combination of external and internal forces with zero resultant (see Figure 3).

2.2.4 Reactionary Forces – Ground Reaction, Joint Reaction, Gliding/Shear/ Friction

A reaction force is a force that acts in the opposite direction to an action force. Friction is the reaction force resulting from surface interaction and adhesion during sliding. Reaction forces and reaction moment are usually the result of the actions of applied forces.

When reaction forces exceed action forces, structure failure can occur and may lead to fracture as well as corrosion. A reaction force is the force applied to a structure when it rests against something. In analyzing a beam structure, it involves calculating what the reaction forces are at the supports due to the forces acting on the beam. A free body diagram of the entire beam can be used to determine the reaction forces.

The beam has vertical and horizontal reaction forces as well as a reaction moment. The horizontal reaction force is zero as long as there are no horizontal applied forces. A cantilever beam with a single support has a reaction force and a reaction moment. The reaction force equals the sum of the applied forces on the beam. If the beam sits on a roller that allows the beam to move freely horizontally, its support has only a vertical reaction force.

When put under compression (or any other type of stress), every material suffers some deformation, even if imperceptible, that causes the average relative positions of its atoms and molecules to change. The deformation may be permanent, or may be reversed when the compression forces disappear. In the latter case, the deformation gives rise to reaction forces that oppose the compression forces, and may eventually balance them. Liquids and gases cannot bear steady uniaxial or biaxial compression — they deform promptly and permanently and do not offer any permanent reaction force.

2.3 Friction

Friction is the resistance to motion of one object moving relative to another. It is not a fundamental force, like gravity or electromagnetism. Instead, scientists believe it is the result of the electromagnetic attraction between charged particles in two touching surfaces.

Scientists began piecing together the laws governing friction in the 1400s, but because the interactions are so complex, characterizing the force of friction in different situations typically requires experiments and can't be derived from equations or laws alone.

For every general rule about friction, there are just as many exceptions. For instance, while two rough surfaces (such as sandpaper) rubbing against each other sometimes have more friction, very smoothly polished materials (such as plates of glass) that have been carefully cleaned of all surface particles may actually stick to each other very strongly.

The frictional force is that force responsible for wear and tear of two bodies. When two bodies in contact, move relative to each other then there exists a force which has a tendency to oppose that movement which is called frictional force.

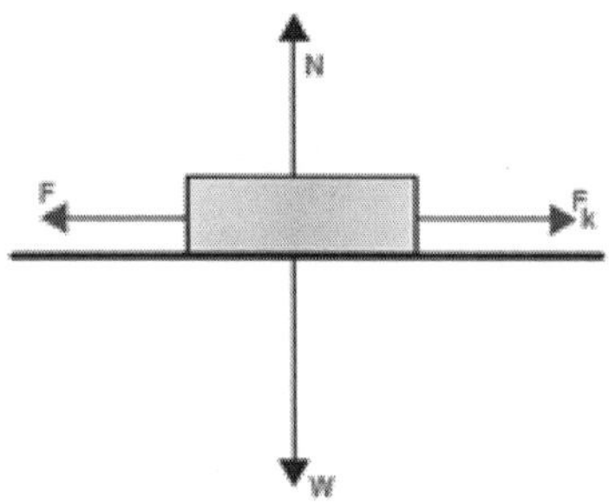

Friction

The friction force depends on two factors:

a. The materials that are in contact. The two materials and the nature of their surfaces. Rougher surfaces have higher coefficient of frictions but to slide apart. This makes sense in terms of a model in which friction is described as arising from chemical bonds between the atoms of the two surfaces at their points of contact: very flat surfaces allow more atoms to come in contact

b. The force pushing the two surfaces together. Pushing the surfaces together causes the more of the asperities to come together and increases the surface area in contact with each other.

2.3.1 Types of friction

There are in general 4 different types of friction.

There are four types of friction namely

- Static friction

- Kinetic friction
- Rolling friction
- Fluid friction

Static Friction

Static friction comes into play when a body is forced to move along a surface but movement does not start. The magnitude of static friction remains equal to the applied external force and the direction is always opposite to the direction of motion. The magnitude of static friction depends upon μs (coefficient of static friction) and N (net normal reaction of the body). Static friction operates between two surfaces that aren't moving relative to each other, while kinetic friction acts between objects in motion.

Kinetic Friction

Kinetic friction denoted as μk comes into play when a body just starts moving along a surface. When external applied force is sufficient to move a body along a surface then the force which opposes this motion is called as kinetic frictional force.

Magnitude of kinetic frictional force $fk = \mu k N$

Where μk is coefficient of kinetic frictional force and N is the net normal reaction on the body. The magnitude of kinetic frictional force is always less than magnitude of static frictional force. When value of applied net external force F is more than fk then body moves with a net acceleration and when these forces are equal then body moves with a constant velocity.

Rolling Friction

Rolling frictional force is a force that slows down the motion of a rolling object. Basically it is a combination of various types of frictional forces at point of contact of wheel and ground or surface. When a hard object moves along a hard surface then static and molecular friction force retards its motion. When soft object moves over a hard surface then its distortion makes it slow down.

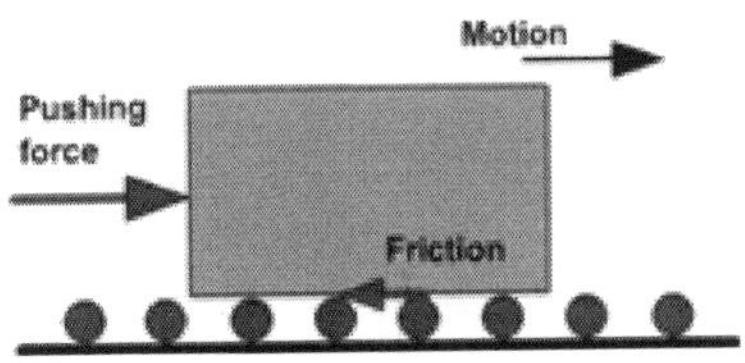

Fig Rolling Friction

Fluid Friction

When a body moves in a fluid or in air then there exists a resistive force which slows down the motion of the body, known as fluid frictional force. A freely falling skydiver feels a drag force due to air which acts in the upward direction or in a direction opposite to skydiver's motion. The magnitude of this drag force increases with increment in the downward velocity of skydiver. At a particular point of time the value of this drag force becomes equal to the driving force and skydiver falls with a constant velocity.

2.4 Addition of Forces: Force Composition

Composition of Forces

If two or more forces are acting simultaneously at a point, the same effect can be produced by a single force of the proper size and direction. This single force, which is equivalent to the action of two or more forces, is called the resultant. Putting component forces together to find the resultant force is called composition of forces.

The vectors representing the forces must be added to find the resultant. Because avector represents both magnitude and direction, the method for adding vectors differs from the procedure used for scalar quantities (quantities having only magnitude and no direction). To find the resultant force when a force of 5 pounds and a force of 10 pounds are applied at a right angle to point A.

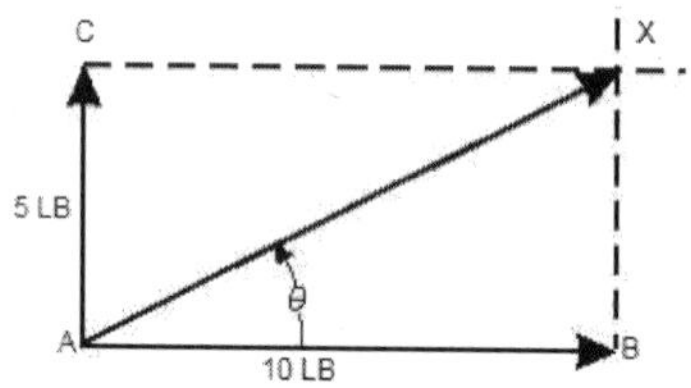

Fig. Composition of two right angle forces

The resultant force may be found as follows: Represent the given forces by vectors AB and AC drawn to a suitable scale. At points B and C draw dashed lines perpendicular to AB and AC, respectively. From point A, draw a line to the point of intersection X, of the dashed lines. Vector AX represents the resultant of the two forces.

Thus, when two mutually perpendicular forces act on a point, the vector representing the resultant force is the diagonal of a rectangle. The length of AX, if measured on the same scale as that for the two original forces, is the resultant force; in this case approximately 11.2 pounds. The angle gives the direction of the resultant force with respect to the horizontal. Mathematically, the resultant force of perpendicular forces can be found by using the Pythagorean Theorem which deals with the solution of right triangles. The formula is C2 = a2 + b2. This states that the hypotenuse, side "C" (our unknown resultant force) squared is equal to the sum of side "a" (one of our known forces) squared and side "b" (another of our known forces) squared.

Let AB and AC represent the two forces drawn accurately to scale. From point C draw a line parallel to AB and from point B draw a line parallel to AC. The lines intersect at point X. The force AX is the resultant of the two forces AC and AB. Note that the two dashed lines and the two given forces make a parallelogram ACXB. Arriving at the resultant in this manner is called the parallelogram method. The resultant force and direction of the resultant is found by measuring the length of line AX and determining the direction of line AX from the figure drawn to scale.

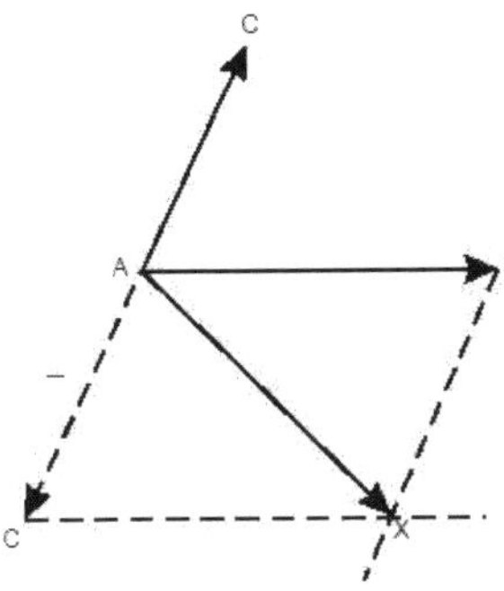

Fig. Parallelogram method of subtracting forces

This method applies to any two forces acting on a point whether they act at right angles or not. Note that the parallelogram becomes a rectangle for forces acting at right angles. With a slight modification, the parallelogram method of addition applies also to the reverse operation of subtraction. Consider the problem of subtracting force AC from AB.

Concurrent Forces

Concurrent forces do not act in one line but they all pass through a common point.

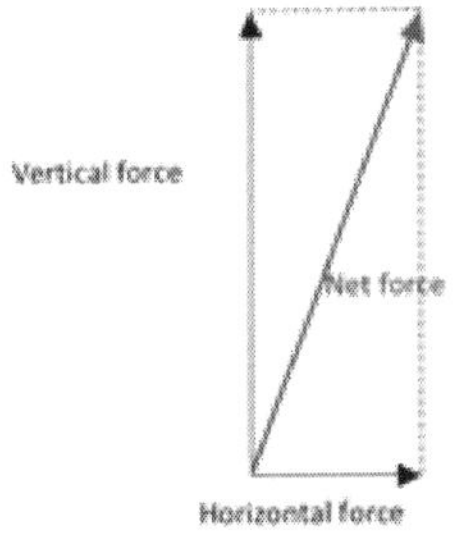

Fig Adding concurrent forces

Resultant force vector is the diagonal of the rectangle created by both horizontal and vertical force and their parallel lines. Resultant force can also be described by the angle between the resultant force and either the vertical or horizontal line. We can arrive at the same result making us of the Pythagoras' theorem. In principle, trigonometry studies triangles and the relationships between their sides and the angles between these sides.

Example

Vertical reaction force of the Earth (normal contact force), acting on a runner's foot, has the magnitude of F_{RA} = 2200 N; friction force is acting backward and its magnitude is F_{RB} = 500 N. What is the direction and the magnitude of the resultant force F_{RC}?

The blue arrow represents the resultant reaction force F_{RC}. Black arrows represent friction component of the reaction force F_{RB} and vertical component of the reaction force F_{RA}, respectively, both acting on the foot.

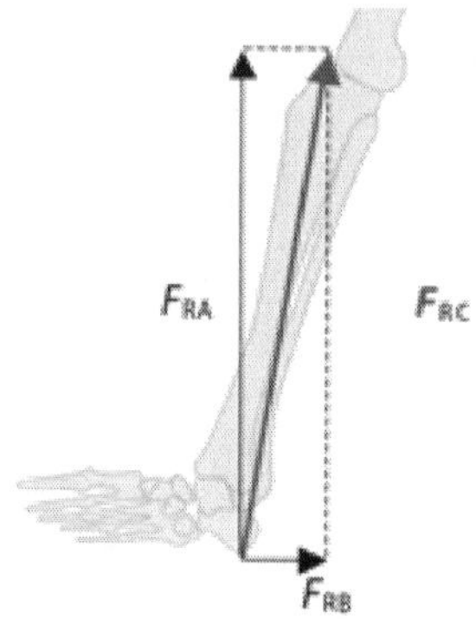

To calculate the resultant force, we will use the Pythagoras' theorem:

$$F_{RA}^2 + F_{RB}^2 = F_{RC}^2$$

$$F_{RC}^2 = 2200^2 + 500^2$$

$$F_{RC} = 2256 \text{ N}.$$

and arctangent function to calculate the angle between the resultant and the horizontal force:

The magnitude of the resultant force F_{RC} is 2256 N and its angle with horizontal line is α = 77,2°.

The magnitude of the resultant force from two perpendicular forces can be calculated with the use of the Pythagoras' theorem, its direction can be calculated with the use of trigonometry.

2.5 Resolution of Forces

2.5.1 Ground Forces Reaction

When we walk, we also produce friction forces on the ground. For example, when the foot hits the ground at heel strike, the friction between the heel and the ground causes a horizontal force to act backwards against the foot:

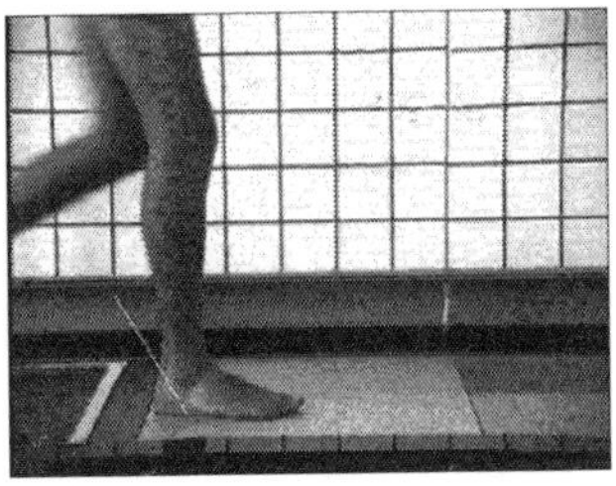

Fig Forces on ground

What will be the effect of this force on the body? As per Newton's equation, F = m.a, it works for horizontal, friction forces as well as for vertical, load forces. The equation tells us that if there is a backwards force on the body, there must also be a backwards acceleration. The force therefore causes a braking action on the body, slowing it down. What about at the end of the stance phase, at toe-off?

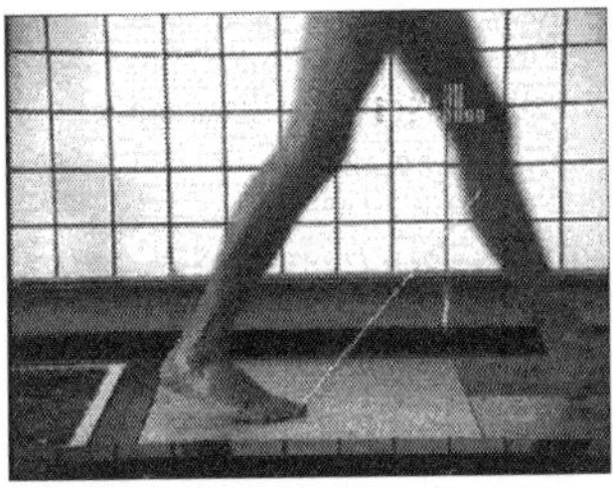

Fig Breaking action of force on the body

Here, the force is tilted over forwards, causing a forwards acceleration on the body, speeding it up into the swing phase. Notice that the ground reaction force is not completely horizontal - it's its tail is still in the air, so there must still be some vertical, load force. In fact, it's very rare to have only a vertical or a horizontal force on the body - usually they both act together as what's known as a ground reaction vector (GRV). We can see how much of each is present at any time by drawing a triangle:

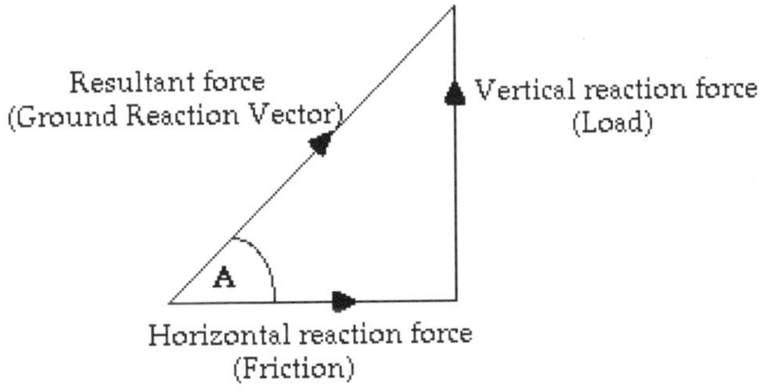

Fig showing ground reaction of vectors

The size of the friction force is represented by the length of the horizontal line (base of the triangle), while the size of the load force is indicated by the length of the vertical line (height of the triangle). The hypotenuse of the triangle is known as the resultant force. It's Its size can be calculated by Pythagoras:

$$\text{Resultant}^2 = \text{Load}^2 + \text{Friction}^2$$

So, it's equal to $\sqrt{(\text{Load}^2 + \text{Friction}^2)}$

It's Its direction is shown by the angle, which can be calculated by trigonometry:

$$\tan A = \text{opposite} / \text{adjacent} = \text{Load} / \text{Friction}$$

This process of breaking a force down into its component parts is called resolving a force, and we will see that it can be a very powerful tool in biomechanics.

Ground Reaction Vector in Normal Gait

We've seen that the GRV is tilted backwards at heel-strike, and forwards at toe-off. What does it do between these times? A plot of the GRV through the gait cycle is called a butterfly diagram because the pattern resembles with wings of butterfly. We can see that the vector swings round from backwards at heel-strike, vertically during the middle of stance, then forwards at toe-off. Notice that It's its height changes, too. It's very small at heel-strike, rises to a maximum soon afterwards, then falls a little during the middle of stance, rises again, then falls before toe-off. This must mean that the vertical reaction force on the body also shows this pattern:

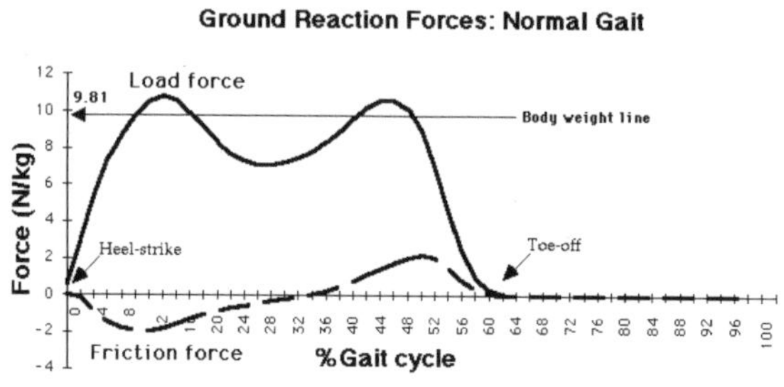

Fig Ground Reaction Vector Forces

The friction force, as we s*aw saw, acts backwards (negative) after heel-strike and forwards (positive) towards toe-off. It changes at some time during the middle of stance - about 36% of the cycle. Notice the scale on the vertical (y) axis. All the forces have been divided by body mass. This makes the graph easier to interpret for people of different body weight. It also shows one very interesting feature. One body weight would be:

$$\text{body mass x gravity / body mass = gravity = 9.81 N/kg}$$

So, we can draw a line at 9.81 to indicate body weight. We call this method of scaling the forces according to body mass normalisation, a technique which is used a lot in biomechanics. Notice that the normalised load force rises above and falls below the body weight line. What does this indicate?

As per Newton's equation, F = m.a, normalised vertical force on the body (in N/kg) must equal the acceleration on the body. If only gravity were acting, the force would be 9.81 N/kg. So, if it's sometimes larger or smaller than 9.81, this must mean that the acceleration on the body is sometimes larger and sometimes smaller than gravity: sometimes adding and sometimes subtracting from the downward pull of gravity:

F = mass x (gravity ± acceleration of body)

2.5.2 Resolution of forces acting on Human Feet

Resolution of forces allows us to analyze causes of motion separately in vertical, mediolateral, and anteroposterior directions. Resolution of forces is very important in sport biomechanics because it allows to better understand human neuromuscular functions in many motor tasks. Amongst examples of motor tasks where it is useful to know individual component forces separately are walking, running, jumping, throwing, postural stability, etc.

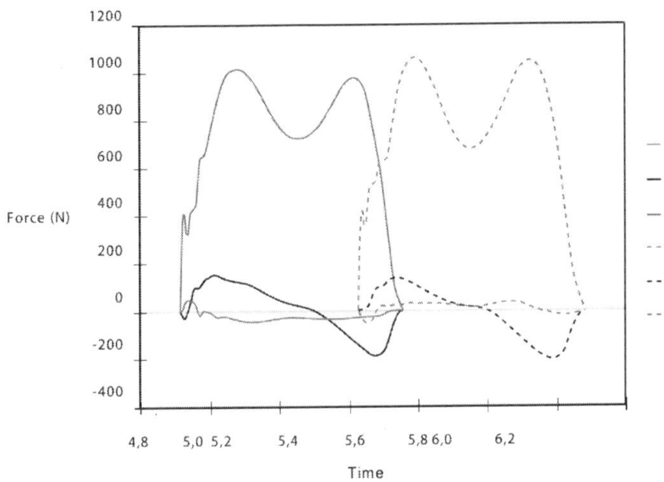

Fig Resolution of forces acting on human feet

Above figure shows the description about resolution of reaction force, acting on human foot while walking that is shown into three component forces. In this:

- Solid line represents components of the reaction force acting on left foot
- Dashed line represents components of the reaction force acting on right foot
- Vertical component forces are marked with blue
- Anteroposterior component forces are marked with black
- Mediolateral component forces are marked with grey

2.6 Static Equilibrium

2.6.1 Statics

According to Newton's first law, a body remains at rest in static equilibrium when its velocity is zero or remains in a state of motion (dynamic equilibrium) when its velocity is other than zero. The study of bodies at rest, as the result of forces acting upon them simultaneously balance each other so that the resultant velocity is zero, is referred to as statics. This balanced state is one of translational equilibrium. That is, during motion, if a body moves in a direction in which a straight line in the body always remains parallel to itself, the motion is called translation. It is a vector quantity measured in feet or meters.

Principles of Statics

Two conditions of equilibrium summarize the principles of statics: (1) For an object to be in linear equilibrium, the total of all X components must equal zero and the sum of all Y components must equal zero. (2) For an object to be in rotatory equilibrium, the total of all torque forces that tend to produce a rotation in one direction (e.g., clockwise) must be counterbalanced by the total of torque forces that tend to produce a rotation in the opposite direction (e.g., counterclockwise).

Free Body Diagram

Clinically, free-body analysis of statics is used whenever traction force is applied; eg, during a manual adjustment of the spine, extremities, or in the use of therapeutic equipment. Other applications are found, for instance, in determining stresses at particular points during activity positions. Freebody analysis is a mathematical technique of utilizing equilibrium equations to determine the internal stresses at a structural point that is being subjected to external load. Such equations can be used to calculate the magnitude of force and moments acting on a vertebral area, for example, at a known position and load.

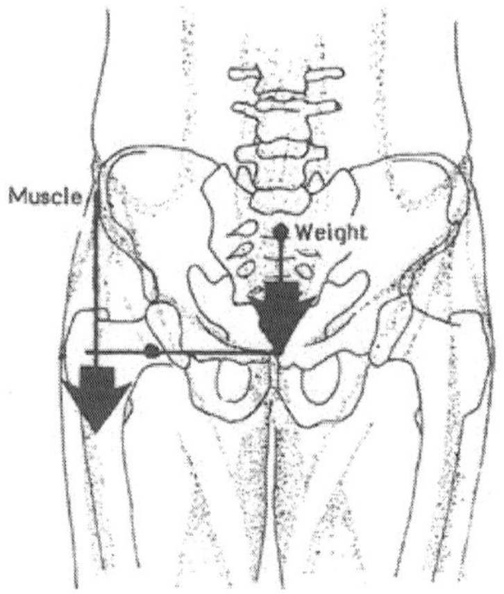

Fig static forces

Equilibrium

While forces of all types may cause subluxations, dislocations, fractures, strains and sprains, etc, the biomechanical mechanisms involved determine the type and extent of the injury produced depending upon the applications of force and its resistance. For example, different applications of force may cause bending, stress, or compression fractures. When the examiner understands how an injury was caused, the tissues involved are more readily located and the extent of injury is more quickly ascertained.

Following is a brief discussion of linear, concurrent, and parallel forces, and their effect upon equilibrium. However, forces acting on a body are sometimes neither linear, nor concurrent, or parallel. General forces always include at least four forces to maintain equilibrium.

Chapter 3
Biomechanical Equilibrium

3.1 Linear Forces

Linear forces are those acting in the same straight line. If two forces are to be in equilibrium in a linear system, the forces must be equal in magnitude and exactly opposite in direction.

3.1.1. Pressure

Pressure refers to how a force is distributed over a surface. Pressure (P) can be defined as the action of a force (F) against some opposing force distributed over an area (A) as in the equation P = F/A, which gives the units of force per unit area such as in pounds per square inch (psi).

Pressure during manipulation

This principle of force is used throughout therapy. In manual adjustment procedures, for example, a patient can withstand a broad palm contact with considerable force without discomfort, yet the same force exerted by a thumb tip or pisiform contact becomes quite painful because the pressure per unit of surface area is now much greater. Thus, whenever pain or skin damage is the priority consideration, the contact forces should be applied over as large an area as possible.

Pressure of support

The same principle must be applied in taping procedures to avoid circulatory and neural impairment, in applying traction slings to distribute force, and in fitting supports if pressure sores are to be avoided. The noxious effects of continuous pressure can be reduced by using somewhat elastic materials such as felt or foam rubber padding as an underlay beneath supports to spread force from prominent bony areas. In large-area supports such as for scoliosis, an underlying water-inflated football bladder has shown to offer automatic pressure distribution and good reciprocity to surface shape. The modern use of air splints during transportation of extremity fractures is another example of this principle applied in health care.

3.1.2 Compression

A pressure always results in a compression stress. Tensile and compression stresses (axial stresses) operate along the axis of a part without altering it. Both of these stresses are measured in newtons. A compression force within the body tends to push substances closer together. When a muscle crossing a joint contracts, it produces a compression force into the joint, and the bones must produce a reactive force to withstand the compression force. Within the spine, the vertebrae and the intervertebral discs are the major compression-carrying components which must support the weight of the body above a particular disc, the initial tension in other ligaments, the additional tension in the muscles and ligaments that are necessary to balance eccentric trunk weight, plus any added external load.

3.1.3 Tension

A pull causes a tensile stress that is an action directly opposed to compression. When tension is applied to connective tissue fibers, the fibers elongate to their physiologic limit somewhat like a stretched rubber band unless a cut or weakness produces a fracture. During torsion (shear) stress, fibers at 45* to the long axis are placed in tension. When a long structure is subjected to bending stress, tension is exhibited in the fibers on the convex side of the curvature.

Motion Tension

Examples of tension are exhibited during all spinal movements. Annulus fibers of the intervertebral discs are placed in tension during disc torsion when the spine is rotated axially, and ligaments posterior to the instantaneous axis of rotation are tensed during spinal flexion. A spinal curvature in any direction involves a constant state of abnormal tension and compression of bones, cartilages, and muscles.

During work, muscles do not maintain a constant tension, length, or move with a constant rate of shortening. The strength of muscle action is affected markedly by the amount of tension in the muscle at the start of movement, the degree of muscle stretch at the beginning of contraction, and the rate at which shortening takes place.

Poisson's Ratio

Similar to a piece of rubber, elastic connective tissue fibers thin during stretch and thicken during compression. In both cases, however, the volume remains constant. The ratio between axial strain in length from compression to transverse strain in diameter from tension is Poisson's ratio.

3.2 Concurrent Forces

In a concurrent force system, as contrasted to a linear system, the forces acting on the body meet at a certain point rather than lie along the same line of action. These forces may be applied to the body from different angles so that their action lines cross either interior or exterior to the body. For example, if two coplanar nonparallel muscles are acting on a bone, a third concurrent force, passing through the point of intersection of the two original muscle forces, must act to maintain equilibrium and avoid rotation.

3.3 Parallel Forces

When parallel forces some distance from each other act upon a body, their forces must be completely nullified by each other if the body is to maintain equilibrium. If forces do not coincide at the same point, such as in a concurrent system, the result is rotation around a stationary axis. A simple form of this action is a force system where the forces lying in the same plane are parallel. Any force acting on an object at a distance from a fixed point tends to rotate the object. Forces producing clockwise rotation are arbitrarily referred to as positive, while counterclockwise forces are termed negative.

The distance from the point of force application (pivot point) to the point of rotation is called the moment arm or lever arm. When a force acts at a dis- tance distance from a pivot point, its effectiveness is determined by both its magnitude and its location.

The tendency of a force to cause rotation about an axis that is equal to the magnitude of the force times the perpendicular distance from the action line of the force to that point is referred to as a moment (torque) of force. Mathematically, it is expressed as moment = force X distance, and its unit of measure is in foot-pounds (ft-lb), kilogram-centimeters (kg-cm), or an equivalent measure.

3.4 Lever Actions

A lever system is a good example of moments developed by coplanar forces. Simply, a lever is a rigid bar turning about an axis. The three components of a lever are the fulcrum upon which the lever turns, the resistance or weight load which is to be moved, and the effort which moves the lever.

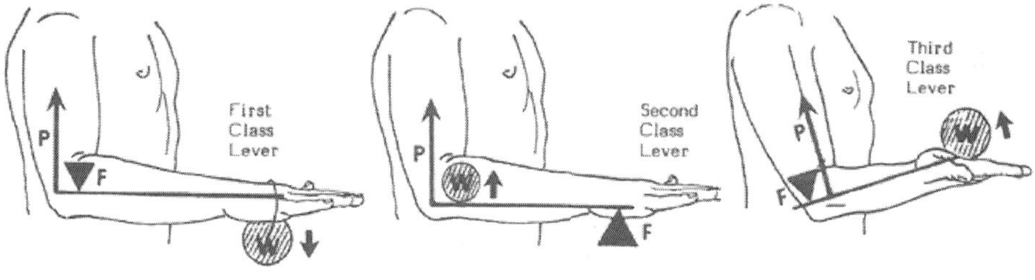

Fig Mechanical arrangements of bone

The articulating surfaces of joints are usually used as fulcrums, the rigid bone shafts extending from axis to axis serve as lever arms, and the source of effort to move the lever arms are the muscles. That is, in the body, we have an effort force and a resisting force acting around a pivot axis (fulcrum) which serves as a supporting force. When a muscle contracts, the bony lever arm to which it is fixed is pivoted about the fulcrum. Work is done when resistance is overcome by a lever system.

Here again is an example that the resistance to be overcome by the body is generally the sum of the load of the lever segment plus the load of the object to be moved.

Advantage

The force developed when a muscle contracts is determined by the mechanical advantage of the levers employed. The position of the attachment of the muscle in relation to the position of the center of the resistance and the position of the fulcrum are important in deciding the mechanical advantage of any lever. Mechanical advantage is the ratio of output force delivered to the input force applied by a mover. If the resistance is placed near the fulcrum, the mechanical advantage is good. If the same resistance is placed further from the fulcrum, the mechanical advantage lessens and a greater effort is required to overcome the resistance.

Mechanical advantage is computed by dividing force into resistance: R/F. If a 150-lb individual wishes to lift a 1200-lb weight by means of a lever, the mechanical advantage required of the lever would be R/F = 1200/150 = 8. Thus, the force arm must be eight times as long as the resistance arm of the lever.

3.4.1 Types of Levers

At one time or another, every bone in the body, acting alone or in combination, acts as a lever, and the human machine offers many examples of first- and third-class levers. Most of the joints of the body can be used as levers of more than one class.

First-Class Levers

An example of a first-class lever where the fulcrum is located between the applied load and the resisting weight is seen in tapping the toes on the floor and in the action of the triceps on the ulnar when the arm is held over the head. Other examples are found (1) when the spleni act to extend the skull across the atlanto-occipital joints and (2) when the soleus tendon force behind and the gravity force in front of the ankle fulcrum are utilized in standing erect. In a first-class lever system, the longer the lever arm, the less force is required to move or resist the applied load.

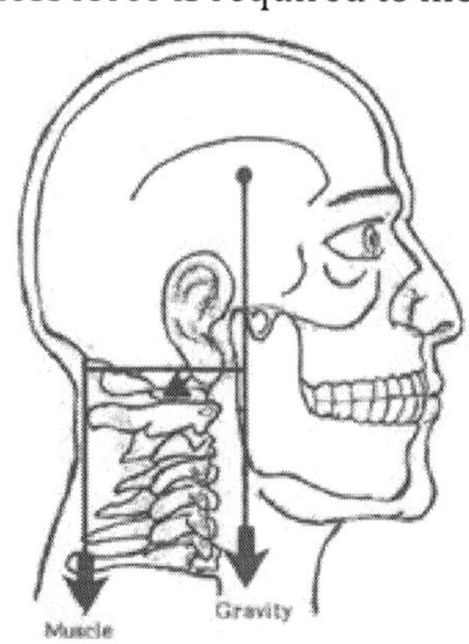

Fig First Class Lever

Second-Class Levers

A second-class lever has the resistance weight located between the applied force and the fulcrum. Since most of the muscles moving each joint are inserted near the joint and the resistance is usually at the long end of the body levers, levers of the first and third classes are the most commonly used in the body. There are no positive second-class levers in the body because the length of the resistance arm of the movement is always greater than that of the effort arm when movement is produced by muscle shortening. Some kinesiologists refer only superficially to the action of the brachioradialis on the forearm during elbow flexion and the action of the calf plantar flexors when one stands on the toes as being second-class lever actions. A true second-class lever system can be demonstrated, however, when muscle tension becomes a resistance to a reversal of joint action caused by an outside force so that a third- class lever system is reversed and becomes a second-class lever system. In such cases, the eccentric muscular contraction is performing negative rather than positive work. An example of this is seen when one stands on tiptoes where the fulcrum is located on the ball of the foot, and force is applied through the gastrocnemius at the ankle to lift body weight.

Third-Class Levers

A third-class lever, where the applied force is always located between the fulcrum and the supported weight (resistance), is seen in the action of the biceps in flexing the forearm. The biceps inserts between the elbow and the hand; thus, when the biceps contracts, the elbow joint serves as a fulcrum. The same mechanism is seen in lifting a weight with the foot.

The resolution of the force of gravity or muscle contraction into components must include a rotary component. Also, the distance from the point of application of the rotary component to the axis of motion in the joint must be considered. The joints which serve as fulcrums also limit the range of movement of the body lever.

3.4.2 Lever Principle

The mechanical advantage derived by the three different lever classes depend on the relative distances between the components, and the levers principle has been established upon this relationship. The resistance is to the effort inversely as the relative distances of the resistance and the effort from the fulcrum. The equation representing the levers principle is: effort X effort arm = resistance X resistance arm. The distance from the resisting force to the fulcrum is called the resistance arm, while the distance from the effort force to the fulcrum is the force arm.

Calculating Effort

The levers principle is used to calculate the effort required to overcome a resistance within a lever system of any class. It is also used to calculate the advantages gained by shifting the position of the resistance, the point of application of effort, and the position of the fulcrum. In the body, the points of application of effort and the position of the fulcrums are somewhat fixed by the anatomic location of the origin or insertion of the muscles in relation to the joints. In some instances, however, changes of posture make effective improvements in mechanical advantage due to shifts in position of the point of application of force and the position of the fulcrum.

3.5 Muscle Force

In the musculoskeletal system, all muscle moment arms are short in proportion to the bony levers they move. The moment arm of the muscle is the perpendicular distance from the muscle's action line of force to the axis of the joint involved. This distance must be used to calculate muscle force rather than the length of the lever arm (distance from the muscle's point of attachment from the joint axis).

3.5.1 Human Potential vs v/s Stability

Each year we see athletic performance draw closer to the limits of human capacity. Understanding the biomechanical principles involved helps us to prevent injury and restore functional integrity. While our lever-like extremities transmit forces and motion at a distance, they also favor musculoskeletal injuries by amplifying forces (usually external, occasionally internal) acting on the body's biomechanical system. Statistics indicate that excessive stress appears greatest on the short arm of first-class levers (eg, elbow, knee).

Linear and Angular Velocity Relationships

The linear velocity of an object at the end of a lever is the product of the lever's length and angular velocity. Thus, an increase in angular velocity and lever length increases the linear velocity because the end of the lever travels further per unit of time. In addition to a longer lever's increased velocity, it takes greater force to move it because the torque exerted by an object at the end of the lever is the product of the length of the lever and its weight.

Wheel and Axial Mechanisms

The wheel-and-axle machine, consisting of a wheel attached to a central axle about which it revolves, works on the lever principle. Rotational force may be applied to either the rim (eg, a steering wheel) or axle (eg, a drive shaft).

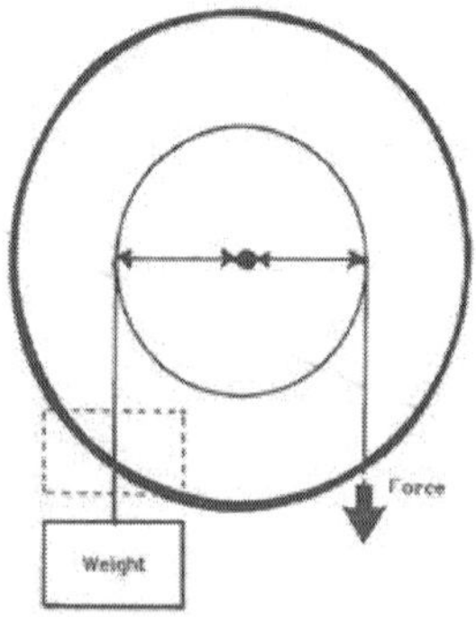

Fig wheel-and-axle machine

Numerous examples of such mechanisms are found within the human machine. For instance, all joint rotation movements in the body involve such a mechanism. On cross section, one readily sees that a long bone of an extremity serves as an axle and its surrounding muscles as a wheel. In the thorax, the rib cage serves as a wheel and the spine as the axle. Here, force is applied to the ribs by the oblique abdominal muscles and to the vertebrae by the deep spinal muscles.

3.6 Pulley Systems

A pulley system may be constructed as either a linear or concurrent force system. A pulley functions as that of a first-class lever with equal arms. Within the body, various fixed pulley systems that may act in any direction are utilized to alter the angle of action on the body by providing resistance, stabilization, and/or to assist movement.

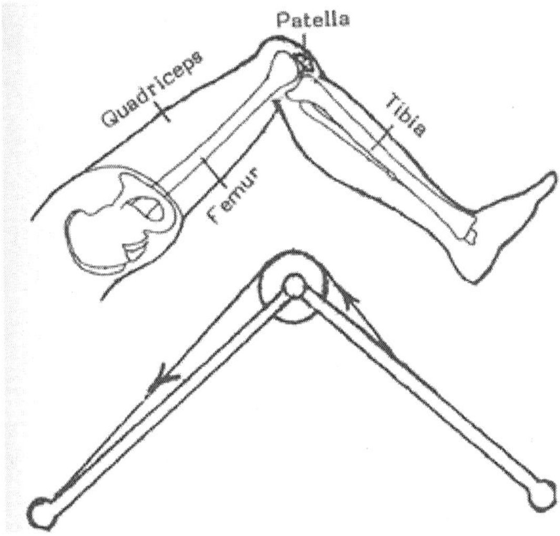

Fig Biomechanical pulley

An example of a fixed single pulley in the body is the patella. The position of the patella changes the direction of quadriceps pull on the tibial tuberosity so as to increase the mechanical advantage. Another example can be found in how the external malleolus of the ankle serves to change the pull of the peroneus longus muscle.

3.6.1 Force Couples

A mechanical couple represents a special case of a pair of parallel forces of equal magnitude that act in opposite directions some distance apart and tend to produce rotation. Two non-collinear parallel forces that are equal in magnitude and opposite in direction have a net force that is zero. While no linear motion would occur in the body, these forces, representing a couple, produce a rotation effect on the body.

The torque (T) of a couple is expressed mathematically as T = force in Newton's X perpendicular distance in meters. Pure rotation cannot take place unless there is a couple.

Applications

Examples of muscle actions that have a couple-like action are exhibited in anterior pelvic tilting by contraction of the lumbar extensors and hip flexors, thoracic rotation on the pelvis by contraction of the latissimus dorsi and contralateral external oblique muscles, and rotation of the head about the axis between C1 and C2 by contraction of the splenius capitis and contralateral sternocleidomastoideus muscles. Scapulohumeral rhythm and the depression action of the rotator cuff muscles on the humerus are two other examples of couple mecha- nisms. An interruption in either of these two couples compromises shoulder girdle motion. Typical side-position lumbar and pelvic adjustments and cervical rotatory adjustments incorporate twisting forces to the spine in opposite directions by action of the adjustor's stabilizing and contact points.

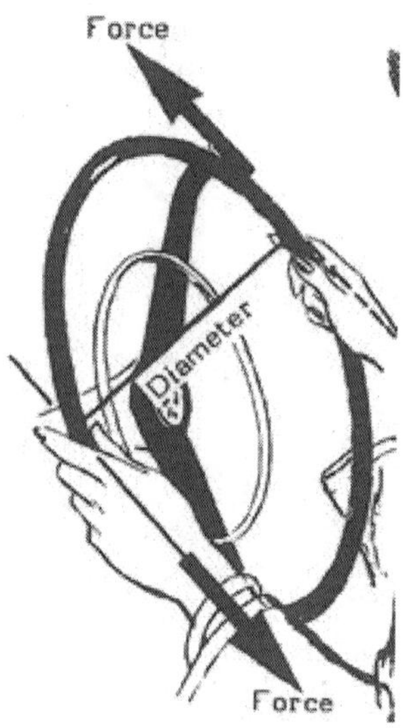

Fig Steering Wheel

A couple consisting of small forces with a large distance between them is just as effective in producing rotation as one where the forces are great and the distance between them is small. For example, in a rotatory adjustment, a lesser force is necessary to produce the same effect when contact is taken well on a transverse process than one taken closer to the spinous process.

3.7 Coupling

A coupling is a device that serves to connect the ends of adjacent parts of objects to produce a phenomenon of consistent association of one motion about an axis with another simultaneous motion about a second axis. In the spine, for example, axial rotation is coupled with lateral bending, vertebral anterior translation is coupled with flexion rotation, lateral scoliotic deformity is coupled with axial rotation to rotate the posterior vertebral elements toward the concavity of the curvature, and vertebral movements toward and from the sagittal plane are coupled with associated rotatory and translatory movements.

3.7.1 Bending

If a load is applied to a relatively long structure that is not directly supported at the point where the load is applied, the resulting deformity is called bending. During bending, the fibers on the concave side of a connective tissue structure are compressed, while those on the convex side are stretched.

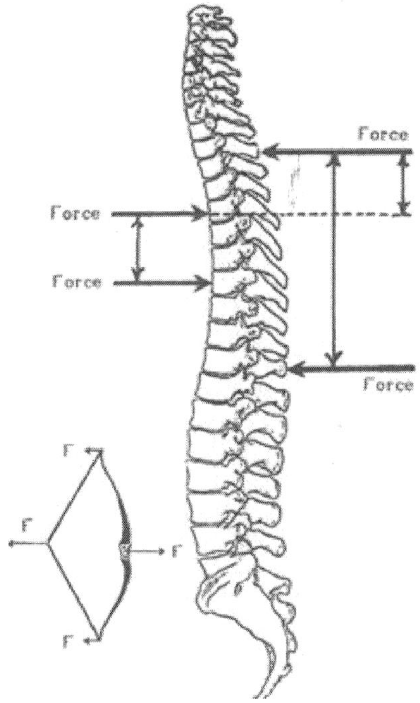

Fig Role of force on spine

Two effects are seen when a force acts on an object. First, the object tends to move in the direction that the force is applied (translation). Second, the force will cause the object to rotate (bending moment).

3.7.2 Bending Moment

A bending moment (torque) is a quantity, usually measured in newton meters, at a point in a structure that is equal to the product of the applied force and the shortest distance from the point to the force direction. For example, think of the trunk of a person sitting as a tree trunk and a laterally abducted arm as a branch of the tree. If a heavy book is placed on an out- stretched hand, the bending moment increases in magnitude from zero at the hand to the maximum at the junction of the tree trunk and branch (shoulder). In the same manner, relatively small weights placed on horizontal limbs apply substantial bending moment at the thoracic and lumbar discs because of the long lever arm. If equilibrium is to be maintained, this stress must be compensated by the bending moment produced by large magnitudes of muscle and ligament force because of the much shorter lever arms of these tissues.

Body weight acting through the center of gravity and the first class lever system of the hip tends to rotate the trunk toward the midline. To maintain equilibrium, this turning effect about the hip from the force of body weight must be resisted by an equal and opposite turning effect produced by the abductors pulling the pelvis downward. The bending moment is the product of force times the perpendicular distance from the force to the center of rotation.

Spinal bending involves the multiple actions of tension, compression, and torsion. The amount of this fiber stress (S) equals the bending moment (B) divided by the sectional moment (I) of inertia times the fiber distance from the neutral axis (Y): $S = B/I \times Y$. If the radius (R) of the curvature is desired, it may be computed by $R = E/B \times I$, where E is the modulus of elasticity of the material.

The moment of inertia of an elliptical object is greatest for bending loads in the direction that is parallel to its major axis. For this reason, the elliptical cross section of a vertebra's pedicle is seen to be most suitable for taking up bending loads in the sagittal plane where such loads are common.

3.7.3 Multipoint Bending

Both three-point and four-point bending occur within the body. Three-point bending is a form of bending where one force is applied to one side of a structure and two forces are applied on the other side. Examples are seen with a seesaw or a drawn bow string. In four-point bending, two transverse forces are applied on one side of a structure and two are applied on the other. If the forces are equal and symmetrical, the structure between the inner two forces is constantly subjected to bending moment.

Neutral Axis

When a long fibrous structure is subjected to bending, the longitudinal line where normal axial stress is zero is referred to as the neutral axis. The plane of the neutral axis is that area situated between the fibers under tension on the convex side of the curvature and the compressed fibers on the concave side. However, there is usually shear stress along the neutral axis resulting from transverse forces even though the tension-compression stress is zero. In cases of torsion stress applied about the neutral axis, the fibers at the neutral axis will have zero shear stress.

3.7.4 Torsion

The mechanical internal moment or couple of restitution which arises in a cord or rod when twisted is referred to as torsion. That is, torsion or torque is the load that is applied by force couples about the long axis of a structure. The moment of torque is the product of a force and its perpendicular distance from the fulcrum. Thus, torque (T) is synonymous with force (F) times the length of the lever arm (a): T = Fa.

If the torques on either side of the fulcrum are equal, the lever is in equilibrium. As mentioned previously, when a lever is in equilibrium, the sum of the moments of force or torques tending to turn it in one direction (e.g., clockwise) about a given point must equal the sum of the moments of the torques tending to turn it in the opposite direction (counterclockwise) about the same point.

3.7.5 Application Principles

Practically all muscles pull obliquely and some pull with a slight twist. The oblique insertion of the pectoral's major into the humerus, for example, causes the humerus to be rotated as well as adducted during contraction of the muscle. In addition, all angles of pull against the bones change with each fraction of a degree of movement. For instance, the angle of pull of the biceps upon the radius changes with each degree of flexion of the elbow.

If torque is applied to the ends of a curved structure, each cross section of the structure is subjected to both torsion and bending forces. Many researchers feel that this principle is responsible for the low-back pain from disc failure because simple axial rotation of the trunk (load) produces severe torsion and bending stress on the lumbosacral disc.

When considering rotational movement, torque fills the same role that force does for motion in a straight line. That is, the magnitude of the moment increases or decreases the angular velocity of an object to produce acceleration or deceleration of the rotation involved.

3.8 Torsion

3.8.1 Torsional Rigidity

Rotatory stiffness is called torsional rigidity in mechanics and represents the torque per unit, measured in Newton meters per radian, of angular deformation. This rotatory stiffness is a characteristic of all body joints.

3.8.2 Area Moment of Inertia

According to newton's first law, the terms mass and inertia can be used interchangeably.

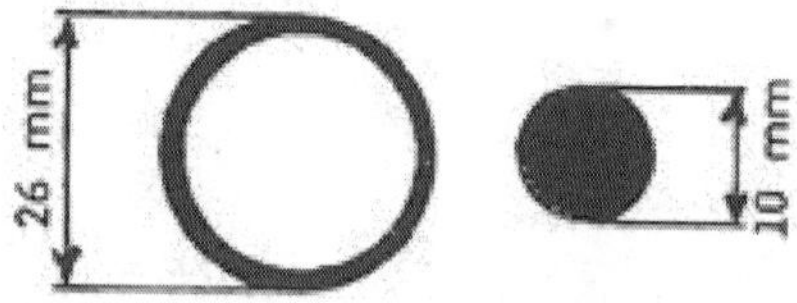

The term "moment of inertia" describes the degree to which a material's shape influences its strength. There are two types: area and polar. The area moment of inertia refers to a measure, in feet or meters to the fourth power, of the distribution of material about its center which determines its bending and torsion strength (i.e., its bending resistance). It is for this factor that a hollow tube such as a long bone is many times stronger in bending and torsion than a solid tube of identical mass.

3.8.3 Polar Moment of Inertia

This is that characteristic of the transverse section of a long object that gives a measure of the distribution of the substance about its axis to increase its torsional strength (resistance). It is also measured in feet or meters to the fourth power. The greater distance a mass is from its axis of torsion, the greater its polar moment of inertia. The reason that the tibia spiral fractures most frequently at the junction of its middle and distal third is because this site has a low polar moment of inertia, even though the cortex is especially thick at this point.

3.8.4 Shear Stress

A force directed against a structure at an angle to its axis that permits one part to slide over the other is called a shearing stress. Typical examples are that of the cervical and thoracic vertebral articular facets and that occurring at the intervertebral disc and vertebral body junction. In shear stress, both articulating parts may be movable with the parts sliding in opposite directions or one part fixed. Shear stress represents the intensity of force parallel to the surface upon which it acts and is measured in newtons/meter524 or pascals. Because bone is weaker in tension than in shear, torsional overstress produces spiral fractures in long bones.

3.8.5 Shear Modulus

Shear modulus is a material property that represents the ratio of a substance's shear stress to its shear strain. As is shear stress, it is measured in newtons/meter. Materials such as rubber and ligamentous tissue with their low modulus of elasticity also have a lower, about 38% lower, modulus of shear.

Chapter 4

Linear Kinetics: Explaining the Causes of Linear Motion

4.1 Introduction

Many treatment approaches used in physical rehabilitation depend on accurate analyses and descriptions of human movement. From the evaluation of these analyses and descriptions, impairments and functional limitations can be identified, diagnoses and prognoses of movement dysfunctions can be formulated, interventions can be planned, and progress can be evaluated. But human movement is often quite complex, frequently being influenced by a dizzying interplay of environmental, psychologic, physiologic, and mechanical factors. Most often, analyzing complex movements is simplified by starting with a basic evaluation of forces acting from within and outside of the body, and studying the effects of these forces on hypothetically rigid body segments.

Newton's laws of motion help to explain the relationship between forces and their effect on individual joints, as well as on the entire body. Even at a basic level of analysis, this information can be used to guide treatment decisions and to understand mechanisms of injury. A simple planar force and torque analysis, for example, provides an estimate of hip joint forces during a straight-leg–raising exercise that may need to be modified in the presence of arthritis or injury. Practicing rehabilitation specialists rarely perform some of the more complex computations described in this chapter; however, understanding the conceptual framework of the computations, appreciating the magnitude of forces that exist within the body, and applying the concepts contained in this chapter are essential to understanding rehabilitation techniques. Such understanding makes clinical work interesting and provides the clinician with a flexible, varied, and rich source for treatment ideas.

Sir Isaac Newton's three laws of motion describe the motion of massive bodies and how they interact. While Newton's laws may seem obvious to us today, more than three centuries ago they were considered revolutionary.

Newton was one of the most influential scientists of all time. His ideas became the basis for modern physics. He built upon ideas put forth from the works of previous scientists including Galileo and Aristotle and was able to prove some ideas that had only been theories in the past. He studied optics, astronomy and math — he invented calculus.

Newton is perhaps best known for his work in studying gravity and the motion of planets. Urged on by astronomer Edmond Halley after admitting he had lost his proof of elliptical orbits a few years prior, Newton published his laws in 1687, in his seminal work "Philosophiæ Naturalis Principia Mathematica" (Mathematical Principles of Natural Philosophy) in which he formalized the description of how massive bodies move under the influence of external forces.

In formulating his three laws, Newton simplified his treatment of massive bodies by considering them to be mathematical points with no size or rotation. This allowed him to ignore factors such as friction, air resistance, temperature, material properties, etc., and concentrate on phenomena that can be described solely in terms- of mass, length and time. Consequently, the three laws cannot be used to describe precisely the behavior of large rigid or deformable objects; however, in many cases they provide suitably accurate approximations.

Newton's laws pertain to the motion of massive bodies in an inertial reference frame, sometimes called a Newtonian reference frame, although Newton himself never described such a reference frame. An inertial reference frame can be described as a 3-dimensional coordinate system that is either stationary or in uniform linear motion., i.e., it is not accelerating or rotating. He found that motion within such an inertial reference frame could be described by three simple laws.

4.2 Newton's First Law of Motion: Law of Inertia

A body at rest will remain at rest, and a body in motion will remain in motion unless it is acted upon by an external force." This simply means that things cannot start, stop, or change direction all by themselves. It takes some force acting on them from the outside to cause such a change. This property of massive bodies to resist changes in their state of motion is sometimes called inertia.

Fig Law of inertia

4.2.1 Law of Inertia

Newton's first law is also called the law of inertia. Inertia is related to the amount of energy required to alter the velocity of a body. The inertia of a body is directly proportional to its mass (i.e., the amount of matter constituting the body).

For example, more energy is required to speed up or slow down a moving 15-pound dumbbell than a 10-pound dumbbell.

Each body has a point, called the center of mass, about which its mass is evenly distributed in all directions. When subjected to gravity, the center of mass of a body closely coincides with its center of gravity. The center of gravity is the point about which the effects of gravity are completely balanced. The center of mass of the human body in the anatomic position lies just anterior to the second sacral vertebra, but the exact position of the center of mass will change as a person changes his or her body position.

In addition to the human body as a whole, each segment, such as the arm or trunk, also has a defined center of mass. In the lower extremity, for example, the major segments include the thigh, shank (lower leg), and foot. Figure shows the center of mass of these segments for the lower extremities of a sprinter, indicated by black circles. The location of the center of mass within each segment remains fixed, approximately at its midpoint. In contrast, however, the location of the center of mass of the entire lower extremity changes with a change in spatial configuration of the segments.

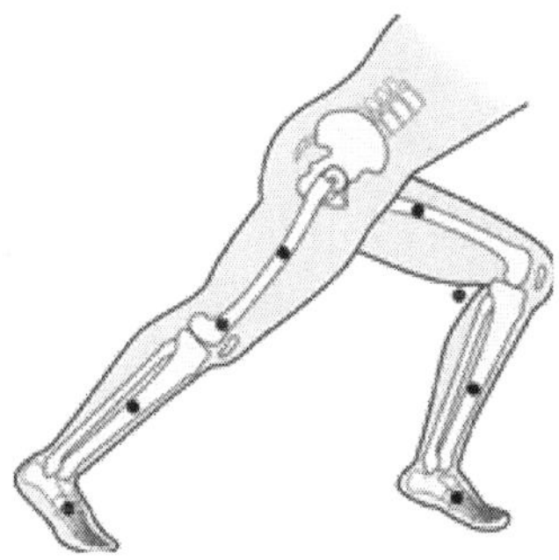

Fig Lower extremities of a sprinter

As shown for the left (flexed) lower extremity, the specific configuration of the segments can displace the center of mass of the lower limb outside the body. Additional information regarding the center of mass of body segments is discussed later in this chapter under the topic of anthropometry.

4.2.2 Mass movement of Inertia

The mass moment of inertia of a body is a quantity that indicates its resistance to a change in angular velocity. Unlike inertia, its linear counterpart, the mass moment of inertia depends not only on the mass of the body, but, perhaps more important, on the distribution of its mass with respect to an axis of rotation. Mass moment of inertia is often indicated by I and is expressed in units of kilograms-meters squared [kg-m^2]. Because most human motion is angular rather than linear, the concept of mass moment of inertia is very relevant and important. Consider again the two positions of the lower extremities of the sprinter in above figure. Within each segment, the individual centers of mass of the thigh, shank, and foot are in the same location in both lower extremities; however, because of the different degrees of knee flexion, the distances of the centers of mass of the shank and foot segments have changed relative to the hip joint. As a consequence the mass moment of inertia of each entire limb changes; the right extended (and "longer") lower extremity has a greater mass moment of inertia than the left.

Another way of conceptualizing the increase is to note that as the knee extends, the center of mass of the entire right lower extremity, depicted by the red circle, moves farther from the hip, thereby increasing its mass moment of inertia. The ability to actively change an entire limb's mass moment of inertia can profoundly affect the muscle forces and joint torques necessary for movement. For example, during the swing phase of running, the entire lower limb functionally shortens by the combined movements of knee flexion and ankle dorsiflexion (in left lower extremity in figure). The lower limb's reduced mass moment of inertia reduces the torque required by the hip muscles to accelerate and decelerate the limb during swing phase. This concept can be readily appreciated during the swing phase while running with the knees held nearly extended (increased I), or almost fully flexed (decreased I).

The concept of mass moment of inertia applies to both rehabilitation and recreational settings. Consider, for example, the design of a prosthesis for the person with a lower limb amputation. The use of lighter components in the foot prosthesis, for example, not only reduces the overall mass (and weight) of the prosthesis, but also results in a change in the distribution of the mass to a more proximal location in the leg. As a result, less resistance is imposed on the remaining limb during the swing phase of gait. The benefit of these lighter components is realized in terms of lessened energy requirements for the person with an amputation. Changing footwear can also make a difference. Consider changes in the mass moment of inertia and resultant required torques for gait when the person changes from wearing a lightweight tennis shoe to a wearing a heavy winter boot.

Athletes often attempt to control the mass moment of inertia of their entire body by altering the position of their individual body segments relative to the axis of rotation. This concept is well illustrated by divers who reduce their moment of inertia in order to successfully complete multiple somersaults while in the air.

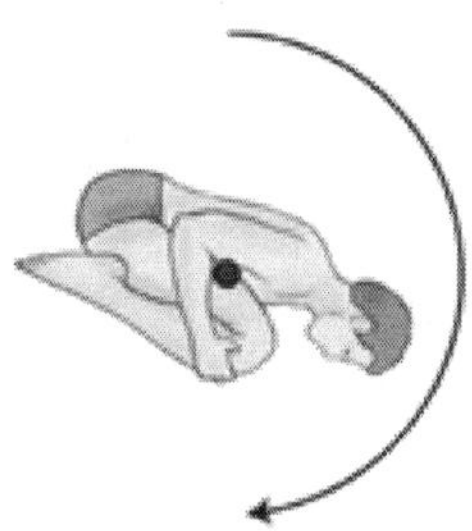

Fig Fastest Angular velocity of athlete

The athlete can assume an extreme "tuck" position by placing the head near the knees, holding the arms and legs tightly together, thereby bringing their segments' centers of gravity closer to the axis of rotation. Based on the principle of "conservation of angular momentum," reducing the body's mass moment of inertia results in an increased angular velocity. Conversely, the athlete could slow the rotation by assuming a "pike" position and increasing the body's moment of inertia, or assuming a "layout" position, which maximizes the body's mass moment of inertia and greatly slows the body's angular velocity.

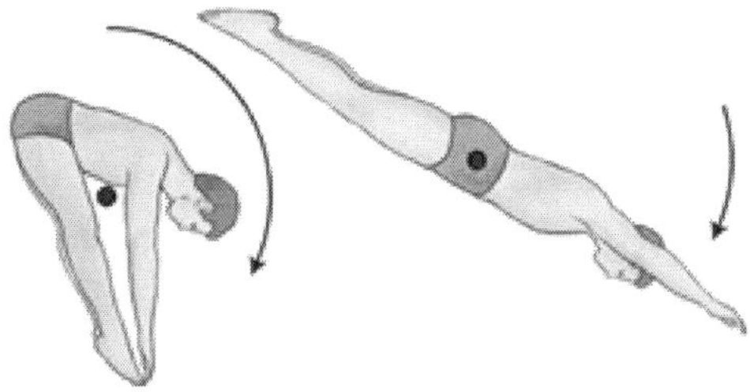

Fig Medium and Slow angular velocities of an athlete

The bat shown in figure is considered to consist of six point masses (m1 to m6), ranging from 0.1 to 0.225 kg, each located 0.135 m from another. During the swing the batter rotates the bat; for illustration purposes the axis of this rotation is positioned as Y_1. If the bat is not sized correctly for the batter, the batter will often "choke up" by shifting his or her grip farther down the bat; again, for illustration purposes, the axis is now positioned as Y_2. The calculations shown in the box demonstrate how the distribution of the mass particles, relative to a given axis of rotation, dramatically affects the mass moment of inertia of the rotating bat. First consider Y_1 as the axis of rotation. The mass moment of inertia of the bat is determined using equation and substituting known values.

$$I = \sum_{i=1}^{n} m_i r_i^2$$

Next, consider Y_2 as the axis of rotation. The important point here is that the mass particles are distributed differently when each axis is considered separately. As seen in the calculations, the mass moment of inertia when considering Y_2 as the axis is 58% of that if Y_1 is the considered axis. This means that the batter could achieve the same angular acceleration with 58% less torque. Or, for the same torque, the bat would accelerate 1.72 times as fast. This is a significant functional advantage gained by choking up on the bat; the bat is easier to swing, although its mass and weight have not changed. The reason for the reduced moment of inertia is that mass points m_2 through m_6 are closer to the Y_2 axis. This is very significant mathematically when one considers that the mass moment of inertia of each point is related to the square of the distance to the axis.

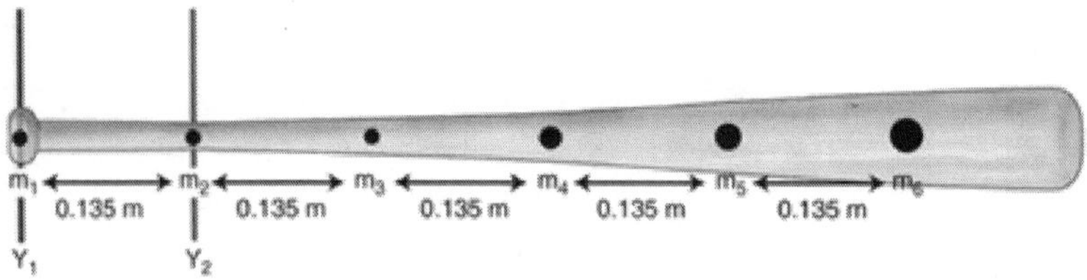

Fig baseball bat rotate around two separate axes of rotation (Y_1, Y_2)

4.3 Conservation of Momentum

The first law says that an object at rest tends to stay at rest, and an object in motion tends to stay in motion, with the same direction and speed. Motion cannot change without an unbalanced force acting. If nothing is happening to you, and nothing does happen, you will never go anywhere. If you're going in a specific direction, unless something happens to you, you will always go in that direction.

Newton's First Law states that an object will remain at rest or in uniform motion in a straight line unless acted upon by an external force. It may be seen as a statement about inertia, that objects will remain in their state of motion unless a force acts to change the motion. Any change in motion involves an acceleration, and then Newton's Second Law applies; in fact, the First Law is just a special case of the Second Law for which the net external force is zero.

Newton's First Law contains implications about the fundamental symmetry of the universe in that a state of motion in a straight line must be just as "natural" as being at rest. If an object is at rest in one frame of reference, it will appear to be moving in a straight line to an observer in a reference frame which is moving by the object. There is no way to say which reference frame is "special", so all constant velocity reference frames must be equivalent.

Conservation of momentum is a fundamental law of physics which states that the momentum of a system is constant if there are no external forces acting on the system. It is embodied in Newton's first law (the law of inertia).

Suppose we have two interacting particles 1 and 2, possibly of different masses. The forces between them are equal and opposite. According to Newton's second law, force is the time rate of change of the momentum, so we conclude that the rate of change of momentum $\mathbf{p}_1$ of particle 1 is equal to minus the rate of change of momentum $\mathbf{p}_2$ of a particle 2,

$$\frac{d\mathbf{p}_1}{dt} = -\frac{d\mathbf{p}_2}{dt}.$$

Now, if the rate of change is always equal and opposite, it follows that the total change in the momentum of particle 1 is equal and opposite of the total change in the momentum of particle 2. That means that if we sum the two momenta the result is zero,

$$\frac{d(\mathbf{p}_1 + \mathbf{p}_2)}{dt} = \mathbf{0}.$$

But the statement that the rate of change of this sum is zero is equivalent to stating that the quantity $\mathbf{p}_1 + \mathbf{p}_2$ is a constant. This sum is called the total momentum of a system, and in general it is the sum of all individuals momenta of each particle in the system.

For electromagnetic radiation,

$$[\text{electromagnetic force}] = [\text{work done}] + [\text{energy density}],$$

so in cgs,

$$\int_S (T \cdot d\mathbf{a})\, d\mathbf{a} = \int_V \mathcal{F}\, dV + \frac{\partial}{\partial t} \int_V \frac{\mathbf{S}}{c^2}\, dV,$$

Where, *T* is the Maxwell stress tensor, $\mathcal{F}$ is the force density, **S** is the Poynting vector, *c* is the speed of light, and $\mathbf{S}/c^2$ is the momentum density.

4.4 Newton's Second Law of Motion: Law of Acceleration

The Second Law of Motion describes what happens to a massive body when it is acted upon by an external force. It states, "The force acting on an object is equal to the mass of that object times its acceleration." This is written in mathematical form as F = ma, where F is force, m is mass, and a, is acceleration. The bold letters indicate that force and acceleration are vector quantities, which means they have both magnitude and direction. The force can be a single force, or it can be the vector sum of more than one force, which is the net force after all the forces are combined.

When a constant force acts on a massive body, it causes it to accelerate, i.e., to change its velocity, at a constant rate. In the simplest case, a force applied to an object at rest causes it to accelerate in the direction of the force. However, if the object is already in motion, or if this situation is viewed from a moving reference frame, that body might appear to speed up, slow down, or change direction depending on the direction of the force and the directions that the object and reference frame are moving relative to each other.

4.4.1 Force (Torque)-Acceleration Relationship

Newton's second law states that the linear acceleration of a body is directly proportional to the force causing it, takes place in the same direction in which the force acts, and is inversely proportional to the mass of the body. Newton's second law generates an equation that relates force (F), mass (m), and acceleration (a) (Equation 4.1). Conceptually, Equation 4.1 defines a force-acceleration relationship. Considered a cause-and-effect relationship, the left side of the equation, force (F), can be regarded as a cause because it represents a pull or push exerted on a body; the right side, m × a, represents the effect of the pull or push. In this equation, ΣF designates the sum of, or net, forces acting on a body. If the sum of the forces acting on a body is zero, acceleration is also zero and the body is in linear equilibrium. As previously discussed, this case is described by Newton's first law. If, however, the net force produces acceleration, the body will accelerate in the direction of the resultant force. In this case, the body is no longer in equilibrium.

F=mxa

The rotary or angular counterpart to Newton's second law states that a torque will cause an angular acceleration of a body around an axis of rotation. Furthermore, the angular acceleration of a body is directly proportional to the torque causing it, takes place in the same rotary direction in which the torque acts, and is inversely proportional to the mass moment of inertia of the body. (The italicized words denote the essential differences between the linear and angular counterparts of this law.) For the rotary condition, Newton's second law generates an equation that relates the torque (T), mass moment of inertia (I), and angular acceleration (α).

$$\Sigma T = I \times \alpha$$

In this equation, ΣT designates the sum of, or net, torques acting to rotate a body. Conceptually, Equation 4.3 defines a torque–angular acceleration relationship. Within the musculoskeletal system, the primary torque producers are muscles. A contracting biceps muscle, for example, produces a net internal flexion torque at the elbow. Neglecting external influences such as gravity, the angular acceleration of the rotating forearm is proportional to the internal torque (i.e., the product of the muscle force and its internal moment arm) but is inversely proportional to the mass moment of inertia of the forearm-and-hand segment. Given a constant internal torque, the forearm-and-hand segment with the smaller mass moment of inertia will achieve a greater angular acceleration than one with a larger mass moment of inertia. (A smaller mass moment of inertia can be achieved by moving a cuff weight from the wrist to the mid-forearm, for example.) Understand that this inertial resistance to the angular acceleration of the limb applies even in the absence of gravity. Because of changes in the mass moment of inertia, less muscular effort will be required to flex the hip with the knee also flexed than with the knee extended.

Torque is expressed in Newton-meters, where 1 Nm = 1 kg-m^2 × radians/sec^2.

4.5 Impulse and Momentum

4.5.1 Impulse

Impulse is simply a measure of the force applied for a specific time. Impulse = force x time and has units Ns (Newton seconds). Impulse is important in sport because many techniques, particularly throwing activities require the performer to apply as large a force possible for as long as possible. The area of a force time graph gives the impulse.

For example when hitting golf ball the player needs to strike the ball as hard as possible (applying a large force) as the ball will fly off in a very short time anyway. The golfer cannot really control how long the force is applied to the ball but they can increase the magnitude of size of the force. However a Discus thrower needs to apply a force to the discus for as long as possible to put the most energy into the implement. The same is true for the Javelin thrower. The longer 'pull' they can apply to the object the faster it will leave the hand.

Examples:

How can you improve an cricketers throwing technique to make them throw a ball further (increase the impulse)? There are two ways they can increase impulse. One is applying a greater force so moving the arm faster which is likely to involve becoming stronger. The other way is to adjust technique so the force is applied over a longer time (or distance). So beginning the throw with the arm fully back and releasing as far forwards with a final flick of the fingers. However, be careful, there is little point increasing the time the force is applied to the ball if the arm is moving slower (i.e. the forces are decreased).

The cricketer applies a force of 20N to a ball for a 0.5 seconds. What is the impulse?

Impulse = force x time

I = 20 x 0.5 = 10Ns

A hammer thrower applies a force of 30N for 1 second doing 1 turn and 30N for 3 seconds doing 4 turns. What is the difference in impulse?

I (1 turn) = 30 x 1 = 30 Ns

I (4 turns) = 30 x 3 = 90 Ns

So we can see that three times more impulse is generated by doing three turns as opposed to four turns. The force is applied to the hammer for longer increasing impulse.

Force time graphs

Force time graphs are often used to illustrate impulse. The area of the graph = impulse.

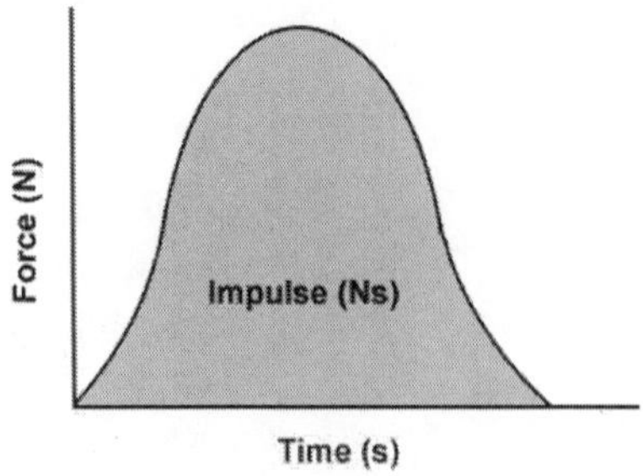

Fig Graph of force V/s time

The graph below shows the force time plot for a vertical jump. When the athlete is standing still only weight from gravity is acting down. As they bend their legs ready to jump the force on the ground is negative and then when they explode upwards into the jump the force becomes positive again.

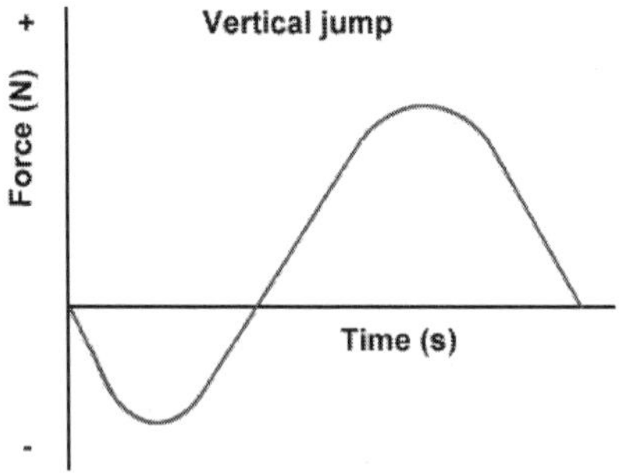

Fig force time graph for vertical jump

4.5.2 Impulse-Momentum Relationship

Additional relationships can be derived from Newton's second law through the broadening and rearranging of equations. One such relationship is the *impulse-momentum relationship.*

Acceleration is the rate of change of velocity ($\Delta v/t$). Substituting this expression for linear acceleration in equation that results in other equations as

$$F = m \times \Delta v/t$$

$$F \times t = m \times \Delta v$$

Application of a linear impulse (force multiplied by time) leads to a change in linear momentum (mass multiplied by a change in linear velocity). The product of mass and velocity on the right side of equation defines the momentum of a moving body. *Momentum* describes the quantity of motion possessed by a body. Momentum is generally represented by the letter *p* and has units of kilogram-meters per second (kg-m/sec). An *impulse* is a force applied over a period of time. The linear momentum of an object such as a moving car is changed by the application of a force over a given time. When a quick change in momentum is required (during an emergency stop, for instance), a very large brake force is applied for a short time. Less brake force for the same time, or the same brake force for even less time, results in a smaller change in momentum. Impulse and momentum are vector quantities. The equation above defines the linear *impulse-momentum relationship.*

The impulse-momentum relationship provides another perspective from which to study human performance, as well as to gain insight into injury mechanisms. At certain locations the body develops mechanisms and structures to lessen peak external load forces. For example, when landing from a jump, peak forces can be reduced throughout the joints of the lower extremities if the impact of the landing is prolonged by more "give" in the muscles—through a lower level and prolonged eccentric activation.

As another example, as the foot contacts the ground during normal gait, the fat pad over the plantar surface of the calcaneus cushions the interaction between the foot and the ground and works to decrease peak reaction forces. Running footwear often augments this function with shock-absorbing outsoles to further cushion the impact of the foot on the ground. Bicycle helmets, rubber or springed flooring, and protective padding are additional examples of equipment designs intended to reduce injuries by increasing the *duration* of impact in order to minimize the peak force of the impact.

Newton's second law involving torque can also apply to the rotary case of the impulse-momentum relationship. Similar to the substitutions and rearrangements for the linear relationship, the angular relationship can be expressed by substitution and rearrangement of equation. Substituting $\Delta\omega/t$ (rate of change in angular velocity) for α (angular acceleration) results in above equations which on rearranging shows angular equivalent of impulse-momentum relationship. Torque and angular momentum are also vector quantities.

$$T = I \times \Delta\omega/t$$

$$T \times t = I \times \Delta\omega$$

Application of an angular impulse (torque multiplied by time) leads to a change in angular momentum (mass moment of inertia times a change in angular velocity).

4.6 Newton's Third Law of Motion: Law of Action-Reaction

Third Law of Motion states, "For every action, there is an equal and opposite reaction." This law describes what happens to a body when it exerts a force on another body. Forces always occur in pairs, so when one body pushes against another, the second body pushes back just as hard. For example, when you push a cart, the cart pushes back against you; when you pull on a rope, the rope pulls back against you; when gravity pulls you down against the ground, the ground pushes up against your feet; and when a rocket ignites its fuel behind it, the expanding exhaust gas pushes on the rocket causing it to accelerate.

If one object is much, much more massive than the other, particularly in the case of the first object being anchored to the Earth, virtually all of the acceleration is imparted to the second object, and the acceleration of the first object can be safely ignored. For instance, if you were to throw a baseball to the west, you would not have to consider that you actually caused the rotation of the Earth to speed up ever so slightly while the ball was in the air. However, if you were standing on roller skates, and you threw a bowling ball forward, you would start moving backward at a noticeable speed.

Newton's third law states that for every action in nature, there is an equal and opposite reaction. This law is demonstrated by the act of walking or running. As your feet exert force down and backward, you are propelled forward and upward. This is known as "ground reaction force."

This force is also observable in the motion of a gondola. As the driver presses his punting pole against the ground beneath the water's surface, he creates a mechanical system that propels the boat forward along the water's surface with a force equal to that which he applied to the ground.

This law implies that every effect one body exerts on another is counteracted by an effect that the second body exerts on the first. The two bodies interact simultaneously, and the consequence is specified by Newton's law of acceleration ($\Sigma F = m \times a$); that is, each body experiences a different effect, and that effect depends on its mass. For example, a person who falls off the roof of a second-story building exerts a force on the ground, and the ground exerts an equal and opposite force on the person. Because of the huge discrepancies in mass between the earth and the person, the effect, or acceleration experienced by the person, is much greater than the effect "experienced" by the ground. As a result, the person may sustain significant injury.

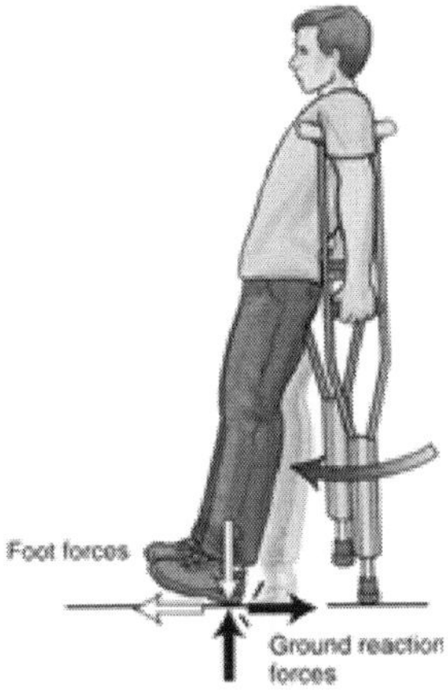

Fig Action of forces produced between ground and foot

Another example of Newton's law of action-reaction is the reaction force provided by the surface on which one is walking or standing. The foot produces a force against the ground, and in accordance with Newton's third law, the ground generates a ground reaction force in the opposite direction but of equal magnitude. The ground reaction force changes in magnitude, direction, and point of application on the foot throughout the stance period of gait. Newton's third law also has an angular equivalent. For example, during an isometric exercise, the internal and external torques are equal and in opposite rotary directions.

4.7 Newton's Law of Universal Gravitation

4.7.1 Introduction

The Law of Universal Gravitation states that every point mass attracts every other point mass in the universe by a force pointing in a straight line between the centers-of-mass of both points, and this force is proportional to the masses of the objects and inversely proportional to their separation This attractive force always points inward, from one point to the other. The Law applies to all objects with masses, big or small. Two big objects can be considered as point-like masses, if the distance between them is very large compared to their sizes or if they are spherically symmetric. For these cases the mass of each object can be represented as a point mass located at its center-of-mass.

While Newton was able to articulate his Law of Universal Gravitation and verify it experimentally, he could only calculate the relative gravitational force in comparison to another force. It wasn't until Henry Cavendish's verification of the gravitational constant that the Law of Universal Gravitation received its final algebraic form:

$F = GxMm/r^2$

Where,

- F represents the force in Newtons
- M and m represent the two masses in kilograms
- r represents the separation in meters
- G represents the gravitational constant, which has a value of $6.674 \cdot 10^{-11} N(m/kg)^2$ as of the magnitude of G, gravitational force is very small unless large masses are involved.

4.7.2 Gravitation force on objects

When considering the gravitational force exerted on an object at a point *inside* or *outside* a uniform spherically symmetric object of radius RR, there are two simple and distinct situations that must be examined: the case of a hollow spherical shell, and that of a solid sphere with uniformly distributed mass.

Case 1: A hollow spherical shell

The gravitational force acting by a spherically symmetric shell upon a point mass *inside* it, is the vector sum of gravitational forces acted by each part of the shell, and this vector sum is equal to zero. That is, a mass mm *within* a spherically symmetric shell of mass MM, will feel no net force (Statement 2 of Shell Theorem).

The net gravitational force that a spherical shell of mass MM exerts on a body *outside* of it, is the vector sum of the gravitational forces acted by each part of the shell on the outside object, which add up to a net force acting as if mass MM is concentrated on a point at the center of the sphere (Statement 1 of Shell Theorem).

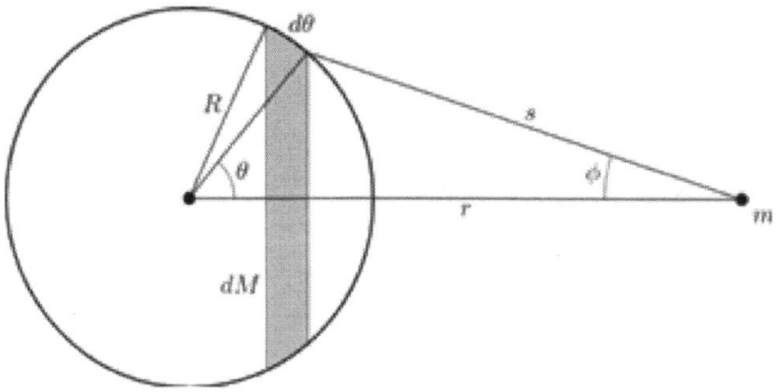

Case 2: A solid, uniform sphere

The second situation we will examine is for a solid, uniform sphere of mass M and radius R, exerting a force on a body of mass mm at a radius d *inside* of it (that is, $d<R$). We can use the results and corollaries of the Shell Theorem to analyze this case. The contribution of all shells of the sphere at a radius (or distance) greater than d from the sphere's center-of-mass can be ignored (see above corollary of the Shell Theorem). Only the mass of the sphere within the desired radius $M<d$(that is the mass of the sphere inside d is relevant, and can be considered as a point mass at the center of the sphere. So, the gravitational force acting upon point mass mm is:

$F=GmM_{<d}/d^2$

Where, it can be shown that $M_{<d}=4/3\pi d^3\rho$

(ρ is the mass density of the sphere and we are assuming that it does not depend on the radius. That is, the sphere's mass is uniformly distributed.) Therefore, combining the above two equations we get:

$F=4/3\pi Gm\rho d$

Which shows that mass mm feels a force that is linearly proportional to its distance d, from the sphere's center of mass.

As in the case of hollow spherical shells, the net gravitational force that a solid sphere of uniformly distributed mass M exerts on a body *outside* of it, is the vector sum of the gravitational forces acted by each shell of the sphere on the outside object. The resulting net gravitational force acts as if mass M is concentrated on a point at the center of the sphere, which is the center of mass, or COM (Statement 1 of Shell Theorem). More generally, this result is true even if the mass M is *not* uniformly distributed, but its density varies radially (as is the case for planets).

Weight of the Earth

When the bodies have spatial extent, gravitational force is calculated by summing the contributions of point masses which constitute them. Newton's law of universal gravitation states that every point mass in the universe attracts every other point mass with a force that is directly proportional to the product of their masses, and inversely proportional to the square of the distance between them.

In modern language, the law states the following: *Every point mass attracts every single other point mass by a force pointing along the line intersecting both points.* The force is proportional to the product of the two masses and inversely proportional to the square of the distance between them:

$F=G\text{x}m_1m_2/r^2$

Where F is the force between the masses, G is the gravitational constant, m1 is the first mass, m2 is the second mass and r is the distance between the centers of the masses.

If the bodies in question have spatial extent (rather than being theoretical point masses), then the gravitational force between them is calculated by summing the contributions of the notional point masses which constitute the bodies. In the limit, as the component point masses become "infinitely small", this entails integrating the force (in vector form, see below) over the extents of the two bodies.

In this way it can be shown that an object with a spherically-symmetric distribution of mass exerts the same gravitational attraction on external bodies as if all the object's mass were concentrated at a point at its center.

For points inside a spherically-symmetric distribution of matter, Newton's Shell theorem can be used to find the gravitational force. The theorem tells us how different parts of the mass distribution affect the gravitational force measured at a point located a distance r_0 from the center of the mass distribution:

1. The portion of the mass that is located at radii $r<r_0$ causes the same force at r_0 as if all of the mass enclosed within a sphere of radius r_0 was concentrated at the center of the mass distribution (as noted above).
2. The portion of the mass that is located at radii $r>r_0$ exerts no net gravitational force at the distance r0r0 from the center. That is, the individual gravitational forces exerted by the elements of the sphere out there, on the point at r_0, cancel each other out.

As a consequence, for example, within a shell of uniform thickness and density there is no net gravitational acceleration anywhere within the hollow sphere. Furthermore, inside a uniform sphere the gravity increases linearly with the distance from the center; the increase due to the additional mass is 1.5 times the decrease due to the larger distance from the center. Thus, if a spherically symmetric body has a uniform core and a uniform mantle with a density that is less than 2/3 of that of the core, then the gravity initially decreases outwardly beyond the boundary, and if the sphere is large enough, further outward the gravity increases again, and eventually it exceeds the gravity at the core/mantle boundary.

4.7.3 Local gravitational acceleration

The law of universal gravitation describes the force between point particles. Yet, it also accurately describes the gravitational force between the Earth and Moon if we consider both bodies to be points with all of their masses concentrated at their centers. The fact that the gravitational force from a spherically symmetric object acts as if all of its mass is concentrated at its center is a property of the inverse square dependence of the law of universal gravitation. If the force depended on distance in any other way, the resulting behavior would be much more complicated. A related property of an inverse square law force is that the net force on a particle inside of a spherically symmetric shell vanishes.

Just as we define an electric field as the electric force per unit charge, we define a gravitational field as the gravitational force per unit mass. The units of a gravitational field are the same units as acceleration, meters per second squared (m/s^2). For a point near the surface of the Earth, we can use Newton's law of universal gravitation to find the local gravitational acceleration, g. If we plug in the mass of the Earth for one of the two masses and the radius of the Earth for the separation between the two masses, we find that g is 9.81 m/s^2. This is the rate at which an object dropped near the Earth's surface will accelerate under the influence of gravity. Its velocity will increase by 9.8 meters per second, each second. Unlike big G, the universal gravitational constant, little g is not a constant. As we move up further from the Earth's surface, g decreases (by 3 parts in 10^5 for each 100 meters of elevation). But it also decreases as we descend down a borehole, because the mass that influences the local gravitational field is no longer that of the entire Earth but rather the total mass within the radius to which we have descended.

Even at constant elevation above sea level, g is not a constant. The Earth's rotation flattens the globe into an oblate spheroid; the radius at the equator is nearly 20 kilometers larger than at the poles, leading to a 0.5 percent larger value for g at the poles than at the equator. Irregular density distributions within the Earth also contribute to variations in g. Scientists can use maps of the gravitational field across the Earth's surface to infer what structures lay below the surface. Just as we define an electric field as the electric force per unit charge, we define a gravitational field as the gravitational force per unit mass. The units of a gravitational field are the same units as acceleration, meters per second squared (m/s^2). For a point near the surface of the Earth, we can use Newton's law of universal gravitation to find the local gravitational acceleration, g. If we plug in the mass of the Earth for one of the two masses and the radius of the Earth for the separation between the two masses, we find that g is 9.81 m/s^2. This is the rate at which an object dropped near the Earth's surface will accelerate under the influence of gravity. Its velocity will increase by 9.8 meters per second, each second. Unlike big G, the universal gravitational constant, little g is not a constant. As we move up further from the Earth's surface, g decreases (by 3 parts in 10^5 for each 100 meters of elevation). But it also decreases as we descend down a borehole, because the mass that influences the local gravitational field is no longer that of the entire Earth but rather the total mass within the radius to which we have descended.

Even at constant elevation above sea level, g is not a constant. The Earth's rotation flattens the globe into an oblate spheroid; the radius at the equator is nearly 20 kilometers larger than at the poles, leading to a 0.5 percent larger value for g at the poles than at the equator. Irregular density distributions within the Earth also contribute to variations in g. Scientists can use maps of the gravitational field across the Earth's surface to infer what structures lay below the surface.

Remember

Linear
1. Time (t)
2. Position
3. Linear displacement (d)
4. Linear velocity (V)= ?d/?t
5. Linear acceleration (A)= ?V/?t
6. Force = mass x acceleration: An imbalancing force causes linear acceleration.

7. Newton's 1st Law - the law of inertia A body at rest remains at rest and a body in motion remains in motion in a straight line unless acted upon by an outside force. <u>Inertia</u> is the property of a body that resists changes in position or linear motion. <u>Mass</u> (M) is a measure of inertia.
8. Newton's 2nd Law - the law of linear acceleration (also known as Law of Momentum) The linear acceleration of an object is directly proportional to the force and inversely proportional to mass. F = M x A, where F is force, M is mass and A is acceleration. Note that F = M x V/t and Ft =MxV Ft is a quantity called impulse and MV is momentum. In other words, the application of force to an object for a period of time causes the object to change its linear momentum.
9. Conservation of linear momentum In any system the linear momentum does not change unless an outside force is applied to the system.
10. Conservation of energy The energy of a system is conserved. Potential energy (PE) = Mgh, it is energy do to position. Kinetic energy (KE) = $1/2MV^{2}$, it is energy do to linear velocity (motion).
11. Newton's 3rd Law - the law of action and reaction For every action (force) there is an equal and opposite reaction (force).

Chapter 5
Work, Power, and Energy

5.1 Introduction

Force is the interaction that causes an object of mass to change its velocity (Force = Mass x Acceleration). In simple terms, it's the maximum energy you can transition into an object. In exercise, force is determined by the max force/torque (rotational force) that a muscle can create.

Force is simply being able to transfer energy into something. The time it takes is negligible because the primary goal is all about the amount, not the speed. Creating force is necessary for ALL types of exercises, which is why "Force" is included in every equation below. Anyone who participates in strength training will ultimately be increasing force production.

Powerlifting is a good example of pure force production. The goal of powerlifting is not to lift a weight really quickly, but to lift at maximum capacity one time. Lifting 400lbs of weight in 3 seconds and lifting 400lbs in 1 second are the same on the scorecard. If you want to carry large things, that's also about force production.

Work is defined as a force applied through a displacement in the direction of the force. Mathematically, work is represented by:

$$W = F * (\cos \theta) * d$$

If the force is in the direction of the displacement, the angle of application is 0°, and the equation reduces to:

$$W = F * d$$

The units of work are Force in Newtons * displacement in meters = Newton-meters (Nm); however, in order to distinguish work from torque, work is assigned a special unit called the Joule (J); 1 Nm = 1 J

Closely related to work is the concept of power. Power is defined as the work done per unit time, or the rate of doing work. Mathematically, power is represented by:

$$P = W / \Delta t$$

Substituting the expression for work into the power equation, we arrive at an alternative expression for power:

$$P = F * d / \Delta t = F * v$$

The units of power are: Work in Joules / time in seconds = Joules/second (J/s). Again, a special unit is defined to take the place of J/s; the unit of power is the Watt.

Energy is defined as the capacity to do work. There are many types of energy, but the type of concern in biomechanics is Mechanical Energy. There are four types of mechanical energy; we will only be concerned with two in this lab - kinetic energy and potential energy.

Kinetic energy is the capacity to do work by virtue of the velocity of an object. It is expressed as:

$$KE = \frac{1}{2}mv^2$$

From the expression, it can be seen that KE can never be negative, as mass is never negative. The units for kinetic energy are: mass in kilograms * (velocity in meters per second)^2 = $kg \cdot m^2/s^2$ = Nm = J

Potential energy is the capacity to do work by virtue of the position above or below a reference position. It is mathematically expressed as:

$$PE = mgh$$

From the definition, PE can be either positive or negative, depending upon position. The units for potential energy are: mass in kilograms * gravitational acceleration in meters per second^2 * height in meters = $kg \cdot m^2/s^2$ = Nm = J

While these concepts are easily defined, difficulties arise when trying to apply these concepts to human motion. Calculating the work and power of human movement is problematic for several reasons. First, it is often difficult to determine the forces that are acting. In such a case, an alternative means of calculating work (and thus power) must be utilized. Based on the definition of energy, the common units of work and energy, and the principle of Conservation of Energy (energy can neither be created nor destroyed, only altered in form), one can equate the work done in a situation to the energy expended (converted to another form). In this respect, we can rewrite the formula for work as follows:

$$W = \Delta E = \Delta KE + \Delta PE$$

$$= \frac{1}{2}m(v_f^2 - v_i^2) + mg\Delta h$$

One might suppose that, with this relationship, calculating the work in a giving given situation might then be very easy. Herein lies a second difficulty in calculating work. Consider, for example, the case of a runner at a constant velocity over level ground. Since velocity is constant, $\Delta KE = 0$; since the height (altitude, elevation, etc.) of the activity is constant, $\Delta PE = 0$ as well. Therefore, an individual running at constant speed over level ground does no work.

5.2 Work

Work[1] means different things to different people because we all have very different jobs that we do every day. But in physics, work means something very specific: 'a change in energy.' This change in energy comes from a force that causes an object to move a certain distance. So in order for work to be done on an object, a force must move that object. You can push on a wall all day, but you're not doing any work on the wall unless you get it moving. In other words, you don't do any work on an object unless you change its energy.

Work is a function of the distance an object moves multiplied by force

(Work = Force x Distance x Cos q)

Basically, it's is how much energy is effectively being used to move something.

While you focus on force production, what you really want is to produce work, because you want the object to move a distance. You don't just want to sit there with weight on your shoulders, resisting against weight if your goal is completing the lift. Nope, the idea of any motion is to move something, not to forever exert force onto an immovable object and get nowhere.

Fig Work done on object

When we move an object, we increase its kinetic energy, which is known as 'energy of motion.' When we bring a moving object to rest, we also do work on the object, but in this case we are decreasing its kinetic energy. Regardless of whether we are increasing or decreasing an object's kinetic energy, the amount of work done is equal to the change in energy. This is an important relationship known as the work-energy theorem.

We can write this statement as an equation that makes it very easy to see the relationship: Work = Δ E, where E is energy, and the Greek letter Delta means 'change in.' So, we read this as: work = change in energy.

5.2.1 Equation related to special cases of Work motion

If the Force is in the Direction of Motion:

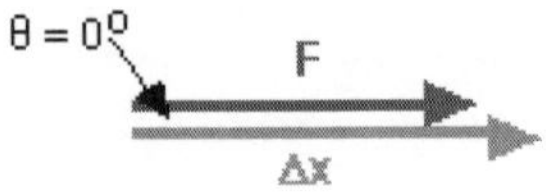

In this case, q = 0°, so:

$$\begin{aligned} Work &= F\,\Delta x \cos\theta \\ &= F\,\Delta x \cos 0^{\circ} \\ &= F\,\Delta x\,(1) \\ &= F\,\Delta x \end{aligned}$$

If the Force is in the Opposite Direction of Motion:

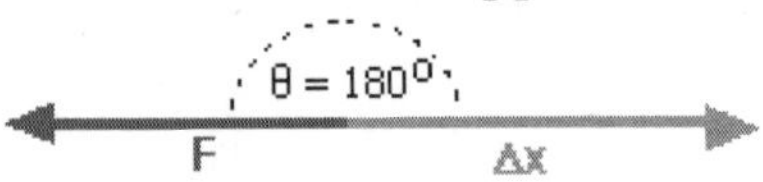

In this case, q = 180°, so:

$$\begin{aligned} Work &= F\,\Delta x \cos\theta \\ &= F\,\Delta x \cos 180^{\circ} \\ &= F\,\Delta x\,(-1) \\ &= -F\,\Delta x \end{aligned}$$

If the Force is Perpendicular to the Direction of Motion:

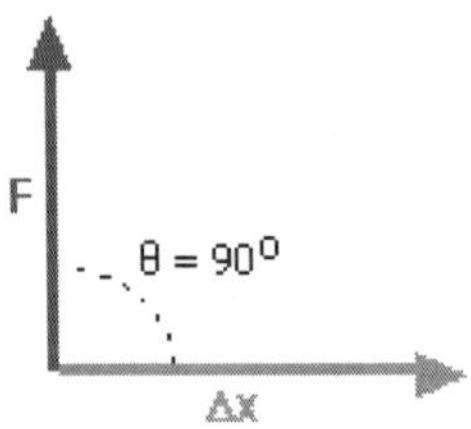

In this case, q = 90°, so:

$$\begin{aligned} Work &= F\,\Delta x \cos\theta \\ &= F\,\Delta x \cos 90^{\circ} \\ &= F\,\Delta x\,(0) \\ &= 0 \end{aligned}$$

If the Object Being Pushed Doesn't Move:

In this case, $\Delta x = 0$, so:

$$\begin{aligned} Work &= F\,\Delta x \cos\theta \\ &= F\,(0) \cos\theta \\ &= 0 \end{aligned}$$

Average Force

Suppose a force is not constant as it does work, but you know (or can estimate) the **average value of the force, F_{ave},** over the distance, x, in which it does work. In this case, the work done by the force is:

Work = $F_{ave}x$ - if the direction of F_{ave} is in the direction of motion

or:

Work = $-F_{ave}x$ - if the direction of F_{ave} is opposite the direction of motion

or, for the mathematically inclined:

$$Work = F_{ave}\,x \cos\theta$$

Area Under the Force vs. Position Graph

If you study the figure shown, you will notice that the work done by a constant force F, acting through a distance Δx can be interpreted as the area under the force vs. position graph for that motion.

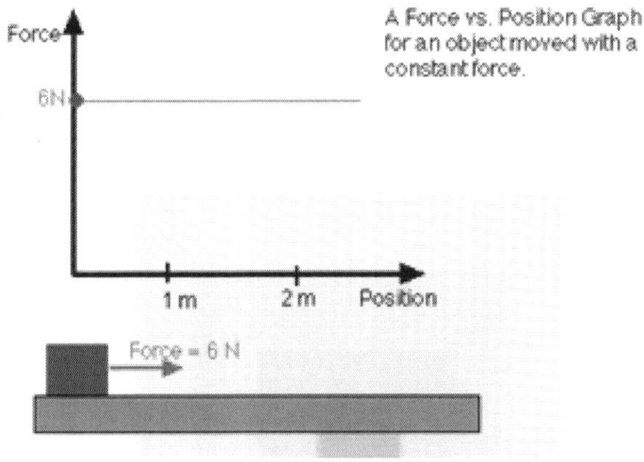

Fig Force Position Graph

Now, work is not always force times distance, but it turns out that **the work done by any force** (or at least the component of the force in the direction of motion) will **always equal the area under the force vs. position graph**.

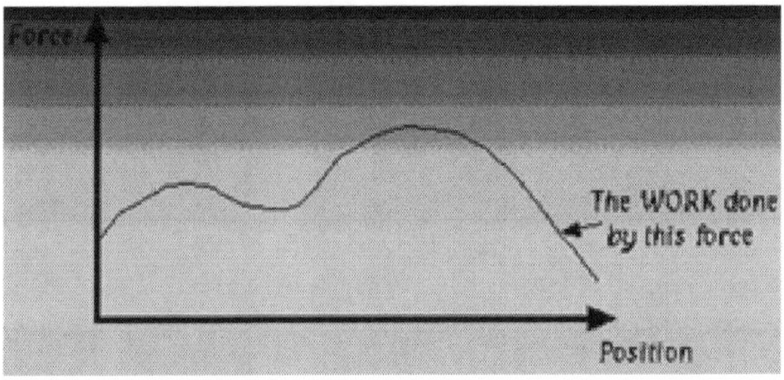

5.3 Energy

In mechanics energy is defined as an ability of a body to perform work. We know many forms of energy: acoustic energy, light energy, chemical energy, nuclear energy, etc. In biomechanics we are most of all interested in mechanical energy, which can have two forms: kinetic energy and potential energy. Energy is denoted by the letter E and the SI unit of energy is joule (J)

There are two forms of mechanical energy:

- kinetic energy, related to motion of a body
- potential energy, related to position of a body in Earth's gravitational field

5.3.1 Kinetic energy

A body in motion has ability to perform work through its motion. Kinetic energy is directly proportional to the square of the body's velocity. Kinetic energy is defined as:

$$E_k = \frac{1}{2}mv^2,$$

Where,

E_k is kinetic energy (J)

m is mass (kg)

v is velocity (m/s)

As we can see in the above definition, to express the magnitude of kinetic energy we have to know the mass and the velocity of the given body (or a human body). What is the kinetic energy of a tennis ball with the mass of 60 g (0,06 kg) and the velocity of 108 km/h (30 m/s)?

$$E_k = \frac{1}{2}mv^2$$

$$E_k = \frac{1}{2}(0,06 \text{ kg})\cdot(30 \text{ m/s})^2$$

$$E_k = 27 \text{ J}$$

Specification of kinetic energy is much simpler than that of mechanical work because to measure the mass and the velocity of a body is easier than to measure forces acting on a body.

5.3.2 Potential energy

Potential energy is the ability of a body to perform work due to its position. Let us mention two types of potential energy: gravitational potential energy and elastic potential energy. *Gravitational potential energy* is, the ability of a body to perform work due to its position in the Earth's gravitational field. Mathematically the potential energy can be expressed as:

$$E_p = mgh.$$

Where,

E_p is potential energy (J)

m is mass (kg)

g is gravitational acceleration (9,81 m/s^2)

h is height (m)

What is the potential energy of a skier with the mass of 90 kg at the start of a downhill run when the altitude difference between the start and the finish is 1000 m?

$$E_p = mgh$$

$$E_p = (90 \text{ kg})\cdot(9{,}81 \text{ m/s}^2)\cdot(1000 \text{ m})$$

$$E_p = 882900 \text{ J}$$

Elastic energy is an ability of a body to perform work due to its being deformed. When, for example, the pole of a pole vaulter is bent, elastic energy is stored in the glass fibre of the vault. The more deformed the pole is, the larger elastic energy is stored in the glass fibre. The amount of elastic energy also depends on rigidity, which is determined by the material qualities of the body.

$$E_p = \frac{1}{2}k\Delta x^2,$$

Where,

E_p is elastic energy (J)

k is rigidity (N/m)

Δx is change to the length of the body (m)

5.4 Work-Energy Relationship

Work W performed by external forces that are acting on a body is the cause of the change of energy ΔE of the body:

W = ΔE

5.4.1 Analysis

Let us have a look at a shot-putter, for example. He used a force of 1206 N to displace the shot by 0,5 m in the direction of the throw. The force used was constant. We will neglect vertical displacement of the shot and thus also its potential energy. Let us imagine that all work was used up to change the kinetic energy of the shot. If we know that the work of 603 J was performed, then the change of the kinetic energy of the shot in the place of throw was 603 J. What was the velocity vodhod of the shot weighing 7,26 kg at the moment of the throw?

$$W = \Delta E$$

$$(630 \text{ J}) = \frac{1}{2}(7{,}26 \text{ kg})\, v_{\text{odhod}}^2$$

$$v_{\text{odhod}} = \sqrt{\frac{2(603 \text{ J})}{7{,}26 \text{ kg}}}$$

$$v_{\text{odhod}} = 12{,}88 \text{ m/s}$$

The initial velocity of the shot at the moment of the throw was 12,88 m/s, as it follows from the relation between work and the change of kinetic energy. The increase in the mechanical energy of the shot equals only the increase in its kinetic energy, because we decided to neglect its potential energy.

Increasing energy by performing work

In sport and physical exercise we often need to maximize kinetic energy of a given body. In order to maximize kinetic energy of human body or sport equipment we must exert the greatest possible force along the longest possible distance.

This way, we can make us of the knowledge of the relation between energy and work to improve our technique in certain sports, especially in athletics. From the relation between impulse of force and momentum we know that for better technique and thus for increasing the velocity of a projectile we need to use greatest possible force for longest possible time. According to the relation between work and energy, the velocity is maximized by the greatest possible force acting along the longest possible distance.

For example shot-putters must throw the shot from a shot-put circle according to the rules, from which they must not step out. Dimensions of the shot-put circle limit the shot-putters' possibility to perform work because they can exert force only along a limited distance. Shot-putters therefore often start their throw by standing on one foot, bent forward over the edge of the shot-put circle, with their back towards the direction of the throw, to maximize the distance along which their force will act on the shot and thus to also maximize the initial velocity of the shot at the moment of the throw as shown.

The longer distance along which the force acts on the shot and the ability to use larger muscle groups thus leads to longer throws and better results. By performing work we can increase not only kinetic but also potential energy of a body.

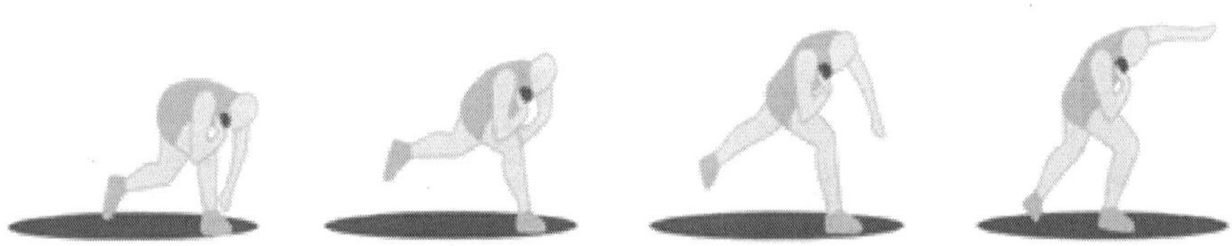

Fig Initial phases of shot put allowing to maximize work performed during throw

Decreasing energy by performing work

The relation between energy and work can also be used to explain techniques of absorbing energy of a body to prevent potential injuries of athletes. This happens mostly in catching projectiles, landing, etc. when negative work is performed. Human muscles also perform negative work when our body lands on the ground.

Energy of a human body during landing is simply given and forces caused by it can lead to injuries if we don't use the correct landing technique. During landing it is important to maximize the distance along which the projectile is decelerating. By making stopping distance longer we make impact forces smaller. We must realize, however, that prolonging the stopping distance by bending our knees deeply, for example, does not necessarily lead to smaller reaction forces in specific joints. The resultant reaction force acting on the floor is smaller but for example in the knee the moment of force can even increase in comparison with landing by bending our knees less and then moving backward.

To decrease impact forces and increase stopping distance we also use various materials: sand (long jump), water (diving), running shoes, boxing gloves, etc.

5.4.2 Law of Conservation of Mechanical Energy

If no work is performed on a body, it follows: change of kinetic energy + change of potential energy + change of elastic energy = 0:

$$\Delta E_k + \Delta E_p + \Delta E_D = 0$$

$$\Delta\left(E_k + E_p + E_D\right) = 0$$

Total mechanical energy of a body is constant unless external forces act on it (other than gravitational force).

$$E_k + E_p + E_D = \text{konst.}$$

This law can be used in studying the motion of projectiles. For example, if we assume that a pole vaulter does not perform any work after taking off, his total mechanical energy at the beginning of the vault is equal to his kinetic energy at the end of his run-up. This kinetic energy is transformed into deformation energy of the pole and subsequently into the increase in potential energy of the athlete. In other words, the faster the pole vaulter runs and the better his pole is able to transform kinetic energy into potential energy through deformation energy, the higher he jumps. Part of energy is of course transformed in other types of energy, for example internal energy of the pole, resulting in heat.

5.5 Power

Power is defined as the rate of work or the rate of energy flow. In certain sports to succeed it is not enough to perform a great amount of work but we must perform a great amount of work in the shortest possible time. In mechanics such ability is described by the quantity called power.

Average power describes the amount of work performed during a period of time. Mathematically this can be expressed as:

$$P = \frac{W}{t},$$

Where,

P is power (W)

W is work (J) a *t* is time period

Instantaneous power is work performed over time period that is approaching zero.

5.5.1 Power measures speed for doing work

Power can also be defined as product of average force and average velocity of displacement of a body in the direction of that force:

$$P = \boldsymbol{F}\boldsymbol{v} = F_t v,$$

where

P is average power (J)

F is average force (N)

F_t is the magnitude of component of average force in direction of body's displacement (N)

v is the magnitude of the average velocity of the body's displacement in the direction of the average force (m/s)

v is the velocity of displacement (m/s)

In complex human motions, maximum output mechanical power is reached with approximately 50 % of maximum force and velocity of a given athlete which means that best gear ratio in cycling is not the top, neither the bottom, but somewhere between the both extremes. The optimum length of step in running is not the longest, neither the shortest but somewhere in between. The best choice of gear ratio, length of step, etc. is the one that allows your muscles to contract with optimum speed and optimum force, which results in maximum mechanical muscle power. In complex motor tasks the resulting power is influenced not only by qualities of individual muscles but also by muscle coordination.

Consider an example of weightlifting, specifically at events of "clean and jerk" and "snatch". The force with which the weightlifter acts on the barbell and the velocity of his motion indicates enormous muscle power (about 3.200 W) but only for a very short time. If the time of motion were longer ??, would he still be able to produce such enormous power? Duration of physical activity substantially influences the ability to produce muscle power. Sprinters are able to maintain power for a relatively short period of time (0-60 s). Power outputs of runners at medium distances are markedly lower than that of sprinters because the duration of their motor activity is increased from one to seven minutes. Marathon runners produce even lower power output because they run for two to four hours. Human metabolic system influences the ability of athletes to produce power.

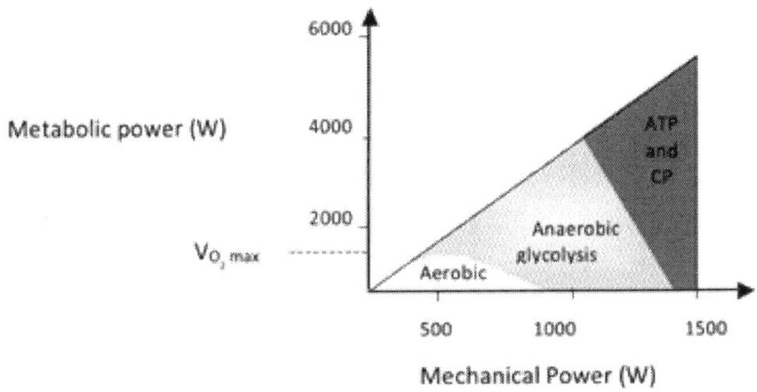

Fig Relation between metabolic and mechanical power, and sources of energy

5.5.2 Power in Joint Kinetics

In joint kinetics, there appears two power measures:

- joint power
- muscle power

The joint power is the scalar product of the net joint force and the joint velocity:

$$P(J) = \mathbf{F} \bullet \mathbf{v} \quad [1]$$

where:

$P(j)$ = joint power

F = net joint force

v = velocity of joint

Note here that power is a scalar: it does not have direction.

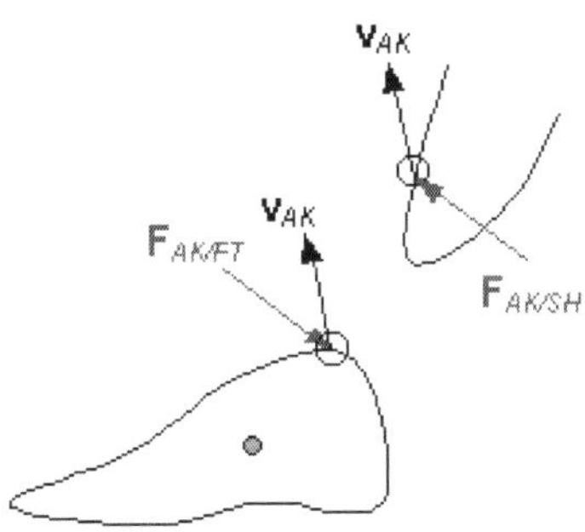

Fig Power of foot

As shown in Figure, the joint power of the foot and shank at the ankle are

$$P(J)_{AK/FT} = \mathbf{F}_{AK/FT} \bullet \mathbf{v}_{AK} \qquad [2]$$

$$\begin{aligned} P(J)_{AK/SH} &= \mathbf{F}_{AK/SH} \bullet \mathbf{v}_{AK} \\ &= -\mathbf{F}_{AK/FT} \bullet \mathbf{v}_{AK} \\ &= -P(J)_{AK/FT} \end{aligned} \qquad [3]$$

Where:

$F_{AK/FT}$ = net joint force acting on foot at ankle

$F_{AK/SH}$ = net joint force acting on shank at ankle

v_{AK} = ankle velocity

Equation [3] shows that foot and the shank show joint powers of the same magnitude with the opposite sign at the ankle. This suggests that the ankle joint only transfers energy from foot to the shank, vice versa. Between the two segments forming a joint, one segment always gains energy at the rate the other loses it, vice versa. Precisely speaking, the joint power is the rate of energy transfer through the joint caused by the linear motion of the joint.

5.5.3 Muscle Power

The muscle power is defined as the scalar product of the joint torque and the segment's angular velocity:

$$P(M) = \mathbf{T} \bullet \boldsymbol{\omega} \qquad [4]$$

Where, P(M) = the muscle power, T = the joint torque, and w = the angular velocity.

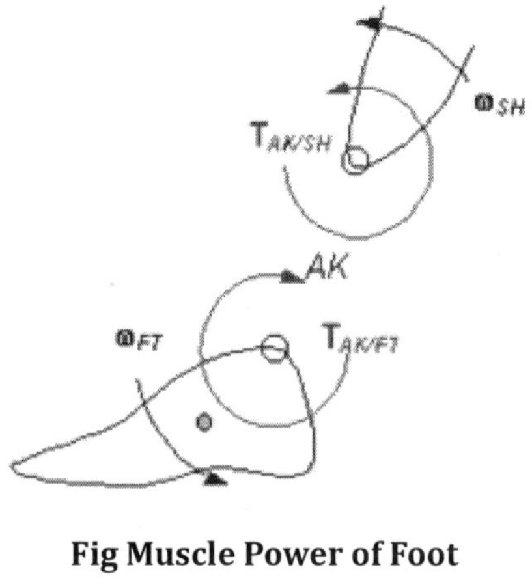

Fig Muscle Power of Foot

For example, the muscle powers of the foot and the shank at the ankle are

$$P(M)_{AK/FT} = \mathbf{T}_{AK/FT} \bullet \boldsymbol{\omega}_{FT} \quad [5]$$

$$\begin{aligned} P(M)_{AK/SH} &= \mathbf{T}_{AK/SH} \bullet \boldsymbol{\omega}_{SH} \\ &= -\mathbf{T}_{AK/FT} \bullet \boldsymbol{\omega}_{SH} \end{aligned} \quad [6]$$

Where,

$T_{AK/FT}$ = the ankle joint torque acting on the foot

$T_{AK/SH}$ = the ankle joint torque acting on the shank

w_{FT} = the angular velocity of the foot

w_{SH} = the angular velocity of the shank

It appears that there is no apparent relationship between the muscle powers of the foot and shank since the angular velocities of the foot and shank can be very different.

At the muscle, two things happen: (1) the muscle transfers energy from one segment it attaches to the other, and (2) the muscle does a work through contraction. The energy transfer happens from one segment to the other and the net change in the energy in the two segments due to the energy transfer must be 0. On the other hand, the muscle can either add (positive work) or drain (negative work) energy to or from both segments at the same time by doing a work. At any rate, sum of the two muscle powers shown in [5] and [6] must yield

$$\begin{aligned} P(M)_{AK} &= P(M)_{AK/FT} + P(M)_{AK/SH} \\ &= \mathbf{T}_{AK/FT} \bullet \left(\boldsymbol{\omega}_{FT} - \boldsymbol{\omega}_{SH}\right) \end{aligned} \quad [7]$$

Where, $P(M)_{AK}$ = the rate of work done by the muscles at the ankle.

If the ankle joint torque and the difference in the angular velocity are in the same direction, $P(M)_{AK}$ becomes positive, suggesting concentric contraction of the muscle while a negative power means eccentric activation of the muscle.

One thing needs to be understood clearly here is, the fact that the works done by the muscles to both segments involved in a joint are not necessarily equal. The actual rates of work done by the muscles to the individual segments depend on the moments of inertia of the segments. The only thing guaranteed here is that the torques exerted by the muscles to the segments are equal as the Third Law of Motion clearly states.

5.5.4 Work Done by Muscles

The amount of work done by muscles at a joint can be computed from the rate of work at the joint through time-integration. For example, the work done by the muscles at the ankle is

$$W_{AK} = \int_{t_0}^{t_1} P(M)_{AK} \cdot dt$$
$$= \int_{t_0}^{t_1} \mathbf{T}_{AK/FT} \bullet (\boldsymbol{\omega}_{FT} - \boldsymbol{\omega}_{SH}) \cdot dt$$

[8]

Where, t_o = the initial time of the movement, t_1 = the final time, W_{AK} = the amount of work done by the muscles at the ankle. [8] Is in fact equal to the area under the muscle power-time curve. Throughout a movement, the muscle power can show both positive and negative phases. Computing the amount of work done in each phase is very important in the sense that it really gives the amount of efforts put by the muscles either concentrically or eccentrically. Integrating the muscle power-time curve for the entire phase is somewhat meaningless since the positive and negative phases cancel each other and one can only get the net work done. The muscles consume energy regardless of the mode of contraction (concentric vs. eccentric) and it is only reasonable to add all the positive and negative works done.

This can be generalized as

$$W_{JOINT}^{(+/-)} = \sum_i \left| \int_{t_0(i)}^{t_1(i)} P(M)_{JOINT} \cdot dt \right|$$

[9]

Where, i = phase number, $t_o(i)$ = the initial time of the i-th phase, $t_1(i)$ = the final time, and $W^{(+/-)}$ = the absolute amount of work done by muscles regardless of the contraction mode. Equation [9] is commonly used to compute the amount of internal work done by muscles and the efficiency of the human machine.

From equations [5] and [6], the overall rate of change in the mechanical energy of the foot and shank are

$$P_{FT} = \{P(J)_{AK/FT} + P(M)_{AK/FT}\} + P(E)_{FT}$$
$$= P_{AK/FT} + P(E)_{FT}$$
$$= \frac{d(KE_{FT})}{dt} + \frac{d(PE_{FT})}{dt} = \frac{d(ME_{FT})}{dt}$$

$$P_{SH} = \{P(J)_{AK/SH} + P(M)_{AK/SH}\} + \{P(J)_{KN/SH} + P(M)_{KN/SH}\}$$
$$= P_{AK/SH} + P_{KN/SH}$$
$$= \frac{d(KE_{SH})}{dt} + \frac{d(PE_{SH})}{dt} = \frac{d(ME_{SH})}{dt}$$

[10]

where $P_{SEGMENT}$ = the overall power (rate of change in the mechanical energy) of the segment, $P_{JOINT/SEGMENT}$ = the energy flow rate into the segment at a joint, P(E) = the power (rate of work) caused by the external forces such as the ground reaction force, PE = the potential energy of the segment, KE = the kinetic energy of the segment, and ME = the mechanical energy of the segment. Equation [10] basically says that the energy flow through the joints and muscles into a segment at both ends (proximal & distal) and the work done by the external forces cause change in the mechanical energy (kinetic energy + potential energy) of the segment.

In the foot, the power caused by the external forces is likely to be zero in most cases since either the foot is stationary on the ground (zero velocity), or the foot is off the ground (zero ground reaction force). If the foot moves while maintaining a contact with the ground such as in slipping or skating, there will be a negative work done by the ground reaction force to the foot. If the main focus is on the work done by the muscles, this power or work is of course not that important.

From equation [10], the change in the mechanical energy of the foot and shank are

$$\Delta ME_{FT} = \int_{t_0}^{t_1} P_{FT} \cdot dt = \int_{t_0}^{t_1} (P_{AK/FT} + P(E)_{FT}) \cdot dt$$
$$= \Delta KE_{FT} + \Delta PE_{FT}$$

$$\Delta ME_{SH} = \int_{t_0}^{t_1} P_{SH} \cdot dt = \int_{t_0}^{t_1} (P_{AK/SH} + P_{KN/SH}) \cdot dt$$
$$= \Delta KE_{SH} + \Delta PE_{SH} \quad [11]$$

5.5.5 Power Curves

For Ankle Joint

A1: a region of negative power, corresponding to eccentric plantar flexor activity at the ankle during midstance and terminal stance.

A2: a region of positive power, corresponding to the concentric burst of propulsive plantar flexor activity during pre-swing.

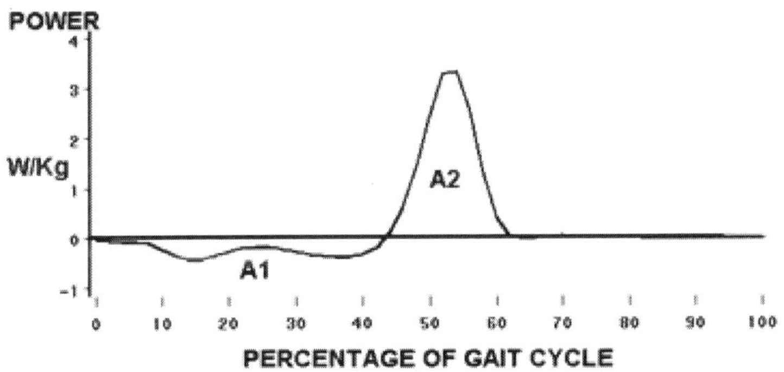

For Knee Joint

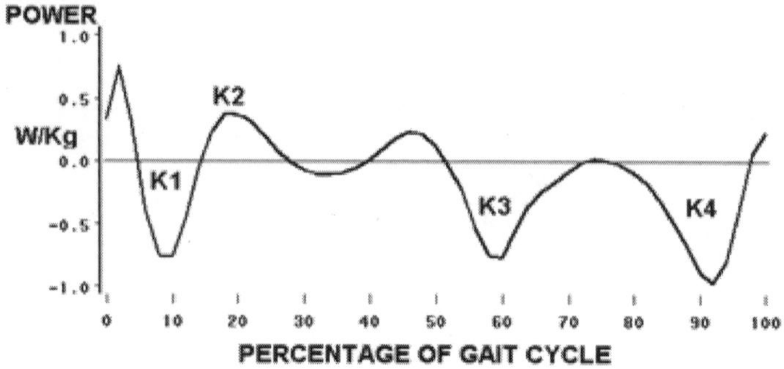

K1: a region of negative power, corresponding to eccentric knee extensor activity at during loading response.

K2: a region of positive power, corresponding to concentric knee extensor activity during mid-distance. This is followed by a period of negligible joint power during that period of time when the ground reaction force stabilizes the knee in extension.

K3: a region of negative power, corresponding to eccentric activity in the rectus femoris during pre-swing. At normal or slightly faster walking speeds, rectus femoris controls knee flexion.

K4: a region of negative power, corresponding to eccentric activity in the hamstrings during terminal swing.

For Hip Joint

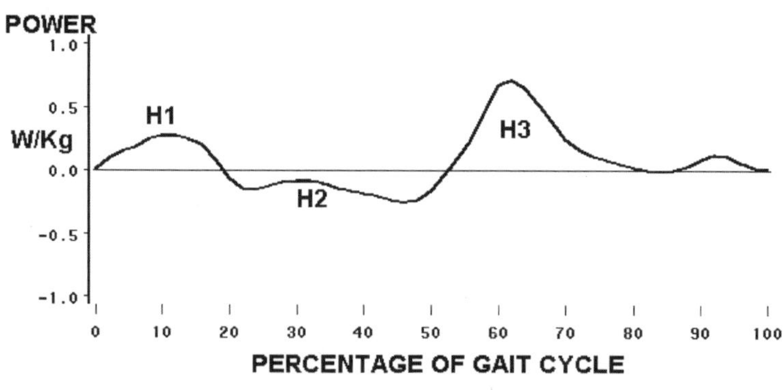

H1: a small region of positive power, not always present, which corresponds to concentric hip extensor activity during loading response.

H2: a region of negative power, corresponding to eccentric hip flexor activity during mid-distance.

H3: a region of positive power, corresponding to concentric activity in the hip flexors during pre-swing and initial swing. Sometimes called "pull off," this is the muscular system's second largest contribution of propulsive power during the gait cycle.

Chapter 6
Torques and Moments of Force

6.1 Torque

We have now identified the rotational equivalent of velocity--namely, angular velocity--and the rotational equivalent of mass--namely, moment of inertia. But, what is the rotational equivalent of force?

Consider a bicycle wheel of radius b which is free to rotate around a perpendicular axis passing through its centre. Suppose that we apply a force $\mathbf{f}$, which is coplanar with the wheel, to a point P lying on its circumference. What is the wheel's subsequent motion?

Let us choose the origin O of our coordinate system to coincide with the pivot point of the wheel--*i.e.*, the point of intersection between the wheel and the axis of rotation. Let $\mathbf{r}$ be the position vector of point P, and let q be the angle subtended between the directions of $\mathbf{r}$ and $\mathbf{f}$.

We can resolve $\mathbf{f}$ into two components--namely, a component $f \cos\theta$ which acts radially, and a component $f \sin\theta$ which acts tangentially. The radial component of $\mathbf{f}$ is canceled out by a reaction at the pivot, since the wheel is assumed to be mounted in such a manner that it can only rotate, and is prevented from displacing sideways. The tangential component of $\mathbf{f}$ causes the wheel to accelerate tangentially. Let v be the instantaneous rotation velocity of the wheel's circumference. Newton's second law of motion, applied to the tangential motion of the wheel, yields

$$M\dot{v} = f \sin\theta;$$

Where M is the mass of the wheel (which is assumed to be concentrated in the wheel's rim).

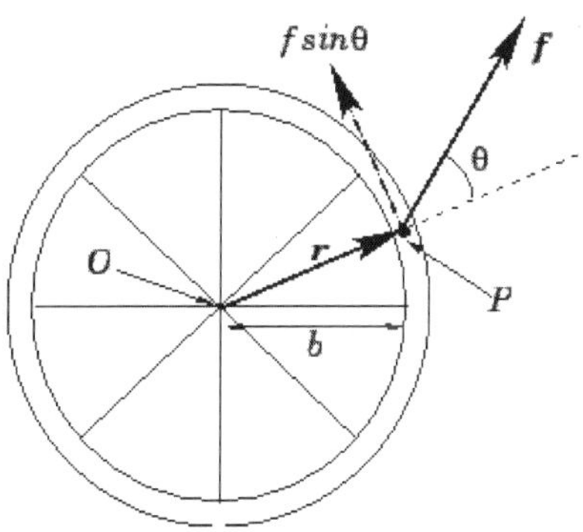

Fig: A rotating bicycle wheel

Let us now convert the above expression into a rotational equation of motion. If ω is the instantaneous angular velocity of the wheel, then the relation between ω and ***V*** is simply

$$v = b\omega.$$

Since the wheel is basically a ring of radius ***b***, rotating about a perpendicular symmetric axis, its moment of inertia is

$$I = M b^2.$$

Combining the previous three equations, we obtain

$$I\dot{\omega} = \tau,$$

Where

$$\tau = f b \sin\theta.$$

Equation is the *angular equation of motion* of the wheel. It relates the wheel's angular velocity, ω, and moment of inertia, ***I***, to a quantity, τ, which is known as the *torque*. Clearly, if ***I*** is analogous to mass, and ω is analogous to velocity, then torque must be analogous to force. In other words, torque is the rotational equivalent of force.

It is clear, from Eq. that a torque is the product of the magnitude of the applied force, ***f***, and some distance $l = b\sin\theta$. The physical interpretation of ***l*** is illustrated in Fig. If can be seen that ***l*** is the perpendicular distance of the line of action of the force from the axis of rotation. We usually refer to this distance as the length of the *lever arm*.

In summary, a torque measures the propensity of a given force to cause the object upon which it acts to twist about a certain axis. The torque, τ, is simply the product of the magnitude of the applied force, ***f***, and the length of the lever arm, ***l***:

$$\tau = f\,l.$$

Of course, this definition makes a lot of sense. We all know that it is far easier to turn a rusty bolt using a long, rather than a short, wrench. Assuming that we exert

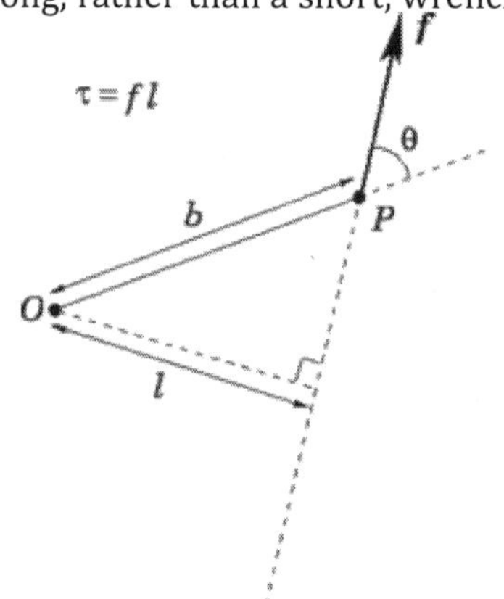

Fig: Definition of the length of the level arm, *l*

the same force on the end of each wrench, the torque we apply to the bolt is larger in the former case, since the perpendicular distance between the line of action of the force and the bolt (*i.e.*, the length of the wrench) is greater.

Since force is a vector quantity, it stands to reason that torque must also be a vector quantity. It follows that Eq. defines the magnitude, τ , of some torque vector, τ . But, what is the direction of this vector? By convention, if a torque is such as to cause the object upon which it acts to twist about a certain axis, then the direction of that torque runs along the direction of the axis in the sense given by the right-hand grip rule. In other words, if the fingers of the right-hand circulate around the axis of rotation in the sense in which the torque twists the object, then the thumb of the right-hand points along the axis in the direction of the torque. It follows that we can rewrite our rotational equation of motion, Eq. in vector form:

$$I\,\frac{d\boldsymbol{\omega}}{dt} = I\,\boldsymbol{\alpha} = \boldsymbol{\tau},$$

Where, $\boldsymbol{\alpha} = d\boldsymbol{\omega}/dt$ is the vector angular acceleration. Note that the direction of α indicates the direction of the rotation axis about which the object accelerates (in the sense given by the right-hand grip rule), whereas the direction of τ indicates the direction of the rotation axis about which the torque attempts to twist the object (in the sense given by the right-hand grip rule). Of course, these two rotation axes are identical.

Although Eq. was derived for the special case of a torque applied to a ring rotating about a perpendicular symmetric axis, it is, nevertheless, completely general. It is important to appreciate that the directions we ascribe to angular velocities, angular accelerations, and torques are merely *conventions*. There is actually no physical motion in the direction of the angular velocity vector--in fact, all of the motion is in the plane perpendicular to this vector. Likewise, there is no physical acceleration in the direction of the angular acceleration vector--again, all of the acceleration is in the plane perpendicular to this vector. Finally, no physical forces act in the direction of the torque vector--in fact, all of the forces act in the plane perpendicular to this vector.

Consider a rigid body which is free to pivot in any direction about some fixed point $\boldsymbol{O}$. Suppose that a force $\boldsymbol{f}$ is applied to the body at some point $\boldsymbol{P}$ whose position vector relative to $\boldsymbol{O}$ is $\boldsymbol{r}$. Let θ be the angle subtended between the directions of $\boldsymbol{r}$ and $\boldsymbol{f}$. What is the vector torque τ acting on the object about an axis passing through the pivot point? The magnitude of this torque is simply

$$\tau = r\ f\ \sin\theta.$$

The conventional direction of the torque is out of the page. Another way of saying this is that the direction of the torque is mutually perpendicular to both $\boldsymbol{r}$ and *f*, in the sense given by the right-hand grip rule when vector ***r*** is rotated onto vector ***f*** (through an angle less than 180° degrees). It follows that we can write

$$\boldsymbol{\tau} = \mathbf{r} \times \mathbf{f}.$$

In other words, the torque exerted by a force acting on a rigid body which pivots about some fixed point is the vector product of the displacement of the point of application of the force from the pivot point with the force itself. Equation specifies both the magnitude of the torque, and the axis of rotation about which the torque twists the body upon which it acts. This axis runs parallel to the direction of τ , and passes through the pivot point.

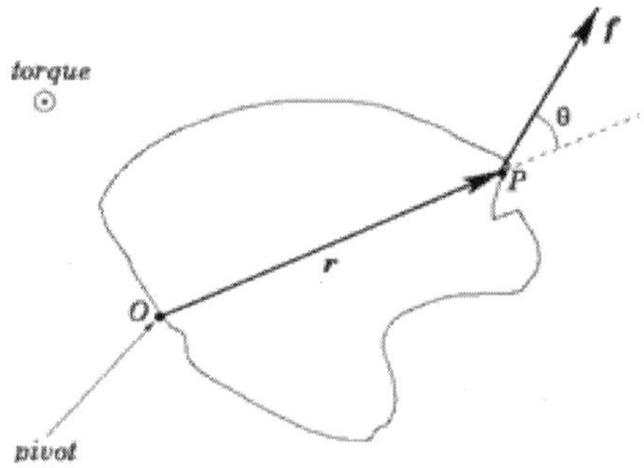

Fig: Torque about a fixed point.

6.1.1 Torque in Biomechanics

Torque is what creates biomechanical movement. It is what creates the movement of the lever system (bones). This is important to understand. Being able to maximize the amount of torque a muscle can generate; will allow for optimal strengthening of that muscle. The greater the torque a muscle can produce, the greater the movement it will produce on the body's levers. If your goal of treatment is to increase movement, you can manipulate the torque variables to maximize the efficiency of the muscles to move the body part. The barbell biceps curl exercise provides a great example of this. It's much harder to move the bar when your elbows are fully extended, compared to when they're at 90°. This is because of the angle-torque relationship. In this relationship, the greatest amount of torque is always when the force is applied at a 90 degree angle to it's its lever.

This concept can also be used with the opposite goal in mind. By adjusting the angle of application and moment arm, you can change the force vector components and increase the amount of compressive force. Increasing the compressive force is often the goal when attempting to maximize stability. The rotator cuff demonstrates this type of force vector when the muscles synergistically contract to create a compressive force couple to stabilize the humeral head in the glenoid fossa.

When attempting to manipulate torque, it is important to realize that the joint ROM does not always correlate with the amount of torque a muscle can create. The angle of insertion is independent of joint ROM. There are many muscles that cross joints that have many different insertions. To determine the force vectors (torque and compression) of a muscle, one must evaluate the line of pull in relation to the lever arms and joint axis.

A common example of the effect of moment arm and angle of application have on torque is the patella's a effect on quadriceps torque. As you can see below, the patella bone increases the angle of application of the quadriceps tendon and therefore moment arm, thus increasing the amount of torque the quadriceps can create. Without the patella most of the quadriceps force would create more of a compression/joint stability moment.

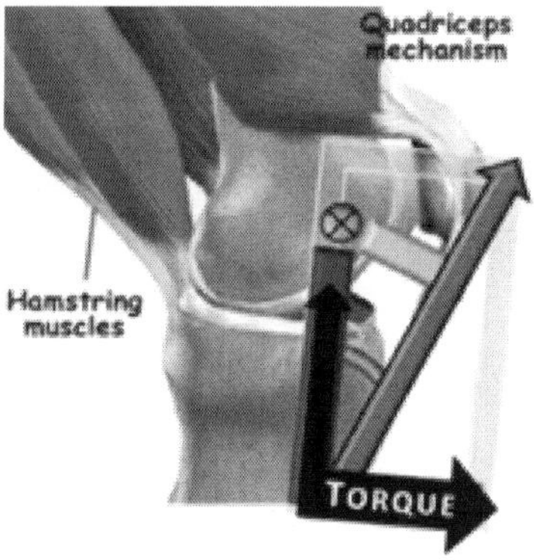

Role of Torque

The patella increases the amount of torque the quadriceps can create due to increased angle of insertion and moment arm

Torque is the driving force for human movement. Being able to manipulate the target muscle torque will allow for a more specific intervention.

- ❖ Moment Arm of a force system is the perpendicular distance from an axis to the line of action of a force.
- ❖ Torque is the ability of a force to cause rotation on a lever
- ❖ Torque is dependent on the amount of force, angle of application of force, and moment arm
- ❖ Joint ROM does not always correlate with the amount of torque a muscle can create
- ❖ Greatest Torque/Moment Arm) = Force applied at 90 degree angle to it's its lever

Torque is the turning effect of a force, calculated as the product of a force (F) and its moment arm (MA), written mathematically as T = F × MA. Before you can sum up the torques, you need to identify the forces that have a moment arm and can create a torque. To do this, go through the problem, identify each force, and give it a label:

- The weight of the dumbbell can be labeled W_D (where Dstands for dumbbell).
- The weight of the forearm/hand segment can be labeled W_S (where S stands for segment).
- The muscle force can be labeled F_M (where M stands for muscle).

Weight is a force that always acts downward. Use a plus sign (+) for the upward direction, and a minus sign (–) for the downward direction. Weights are applied at the center of gravity of a body, and the location of the center of gravity for both the segment and the dumbbell weight are given.

Consider the table where you list items for calculating torques:

Force	Magnitude and Direction	Moment Arm (MA)	Torque (T = F × MA)	Torque Name
WD	-500 N	0.34 m	–170.0 Nm	TD
WS	–17 N	0.23 m	-3.9 Nm	TS
FM	Unknown, to be solved	0.05 m	Unknown	TM

It's important that weights be listed as negative forces. The moment arm for each force is on the same side of the elbow joint axis, so set them all as positive. The moment arms for the segment weight and the dumbbell weight are the distance of each center of gravity from the elbow axis because the forearm/hand is in the horizontal position.

The torque created by each force is calculated as the product of the force and moment arm. The weights (segment and dumbbell) create negative torques, and it's important to list the direction, as well as the magnitude of the torque in the table.

Next, use the equation $\Sigma T = 0$ to solve for the torque created by the muscle (T_M). To do this, expand the equation to list all the torques, like this:

$$T_M + T_D + T_S = 0$$

Now, isolate for the unknown muscle torque:

$$T_M = -T_D - T_S$$

Fill in the known values from the table you created and solve:

$$T_M = -(-170 \text{ Nm}) - (-3.9 \text{ Nm}) = 173.9 \text{ Nm}$$

The muscle must create a torque of 173.9 Nm, opposite in direction to the torques created by segment and dumbbell weights, to prevent angular acceleration.

The last step is to calculate the muscle force (F_M), using the following equation:

$$T_M = F_M \times MA_M$$

Isolate for F_M, making the equation:

$$F_M = \frac{T_M}{MA_M} = \frac{173.9 \text{ Nm}}{0.05 \text{ m}} = 3,478 \text{ N}$$

6.2 Forces in Equilibrium

A body is said to be in equilibrium if it is at rest or moving with uniform velocity. In other words if the linear and angular acceleration of a body are zero, the body is said to be in equilibrium. Or we can say that when two or more forces act on a body such that their resultant or combining effect on the body is zero and the body retains its state of rest or of uniform motion then the body is said to be in equilibrium.

With respect to the state of a body, equilibrium may be divided into two categories:

1. Static equilibrium.
2. Dynamic equilibrium.

In a free body diagram, where two athletes are standing on the ground and F_R is the sum of reaction forces acting on both legs and F_G is gravitational force acting on athlete's body.

Fig free body diagram of athlete standing on the ground

If a simple situation of an athlete standing on the ground as shown appears, then there are two forces acting on the athlete. First force is non-contact gravitational force where such force acting on the athlete, makes the athlete in moving downwards with the acceleration of 9,81 ms^{-2}. As this is obviously not the case, there must exist another force causing the state of static equilibrium, so that the athlete is at rest. This is the reaction force of the ground acting upward on the athlete's feet.

6.2.1 Static Analysis

Mathematically the situation of a standing person can be described as follows:

$\Sigma F = 0$

This is the equation of static equilibrium. The expression ΣF represents the resultant external force, in other words a vector sum of forces acting on a standing person; **o** is zero vector (0, 0, 0).

If forces acting on a person lie in one line, we can simply add them up. Their upward direction is marked as positive, their downward direction as negative.

In the case of a standing person the reaction force of the ground acts upward on the human body and, in line with our convention, has positive direction. However, its magnitude is unknown to us. Gravitational force acts downward and therefore has negative direction.

If only two forces act on a body in the state of either static or dynamic equilibrium, they have equal magnitude but opposite direction.

Weightlifter with the weight of 70 kg has lifted a barbell with the weight of 90 kg and is holding it above his head as shown in figure. As long as he is holding the barbell, both bodies (weightlifter and barbell) are in static equilibrium. What is the force that must act on the weightlifter's feet to keep him in static equilibrium?

Fig Free body diagram – weightlifter with a barbell

If we consider the case of both the bodies (weightlifter and barbell) together, the solution may be quite simple. We sum up the weights of both bodies. The resultant reaction force acting on the weightlifter's feet is then the total of these two gravitational forces, but with opposite (i.e. upward) direction. The force *F***R** equals 1570 N, i.e. the weight of the weightlifter plus the weight of the barbell, and acts vertically upward.

In the case that the forces acting on a body are not in one line and their directions are neither vertical, nor horizontal, we have to resolve these forces into their horizontal and vertical components. Then we solve static equilibrium equations, separately for vertical and horizontal directions. After that we will calculate the magnitude of the resultant force with the use of the Pythagoras' theorem and its direction with the use of trigonometry.

6.2.2 Force vectors

A force vector is the direction of a force. On the page covering the essentials of movement mechanics we talked about the line of force of gravity. There are many force vectors at play when we lift an object. We have gravity, the force of friction on the object, ground reaction forces, muscle forces (generated in the line existing between the origin and insertion of the muscle) and forces of momentum and so on.

The outcome of a movement is the sum of all vectors and their respective forces.

To put this in an exercise context, on punching a boxing bag, following forces and vectors would be in play

- ❖ The ground reaction forces generated from my legs to start the punch
- ❖ The internal forces of every muscle involved from the foot, legs, trunk and arm
- ❖ The momentum of my arm
- ❖ The inertia of the bag

The main thing to do when analysing any exercise is to work out the major forces, how the body is likely to manage them, and how the body will stabilise the joints involved in the movement. Once you have done this you will know:

1. What muscles will be worked during the movement
2. Whether the posture involved in the movement is practical and one that you would want to encourage
3. What muscles will be worked as stabilisers
4. Where the exercise technique would go wrong if a person was to technically fail at this exercise

Back Squat

As an example we can look at squatting, initially with the back squat:

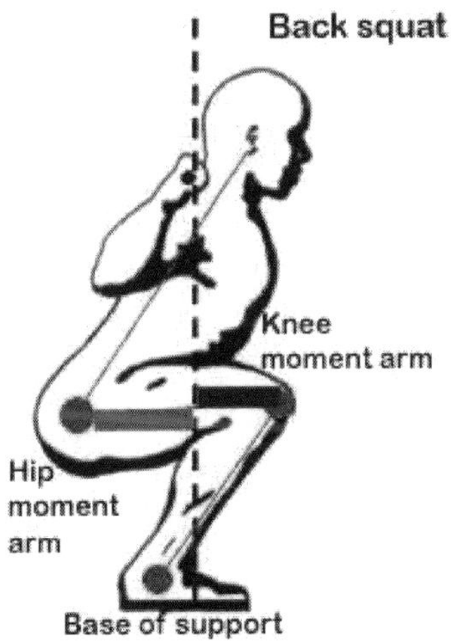

Fig Back Squat

1. In this squat you can see the moment arms around the hip and knee – they measure from the joint axis to the line of force. In this bar position the moment arm around the hip is slightly longer than that around the knee. This means the hip extensors (gluteal muscles) will be doing slightly more work than the knee extensors (quadriceps) in terms of the force they'll need to generate to overcome the load.
2. The posture here is good and is to be encouraged
3. The trunk, knee and hip stabilisers will be worked significantly, especially as more load is added.
4. This squat will most likely be lost from the core, resulting in the lumbar spine rounding and the bar dropping forward, so cueing of 'chest up', 'tummy tight', 'push through the hips and up' will help. As with all squats knee alignment, if lost, can cause counter rotation in the lumbar so cueing knee position as your clients fatigues is also important.

Front Squat

Now let's compare the back squat with the front squat:

Fig Front Squat

1. In this squat you can see the moment arms have changed slightly. They are now about equal meaning the loads around the hip and knee will be similar. This means the hip extensors and knee extensors will need to generate similar amounts of force throughout the movement.
2. This posture is also good and should be encouraged.
3. The shoulder, trunk, hip and knee stabilisers will be worked significantly, as in the back squat.
4. This squat will also be lost from the core or from the shoulder girdle due to the positioning of the bar, so cueing of 'elbows up', 'chest out' and 'tummy

tight' will help. As with all squats knee alignment, if lost, can cause counter rotation in the lumbar so cueing knee position during fatigue is again also important.

Low bar back squat

Now let's look at some squat variations, stating with the low bar back squat:

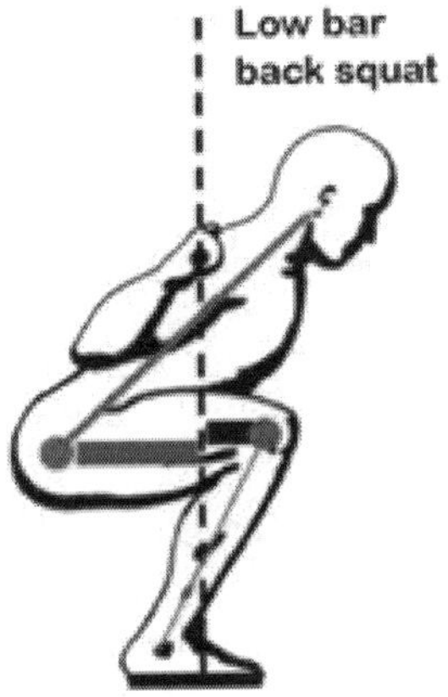

Fig low bar back squat

1. In this squat you can see the moment arm around the hip is at least twice as long as the moment arm around the knee. This low bar position is the position power lifters tend to use as it involves the hip extensors a lot more than the knee extensors and the hip extensors are able to produce more force than the knee extensors and act around a joint with better articulation (deep ball and socket versus the knee which is a shallow joint). As the hip extensors are able to produce more force than the knee extensors then this squat variation enables heavier loads to be lifted.
2. The posture wouldn't be encouraged except where needed for performance in competition (e.g. powerlifting or other sports where this position is required under load) as huge loads are placed on the lumbar spine as well as the hips, and these areas need to be progressively conditioned to withstand these forces. This is an 'advanced' lifters posture.
3. The hip and trunk stabilisers will be worked significantly with this variation.
4. This squat will be lost from the core, resulting in lumbar rounding so cueing 'tummy tight', 'drive up and through', 'head still' will help. As with all squats knee alignment must be maintained.

Finally to understand force vectors a little more, let's take a look at the smith machine squat with feet slightly out in front and then the Swiss ball squat which are commonly prescribed squat variations in fitness clubs.

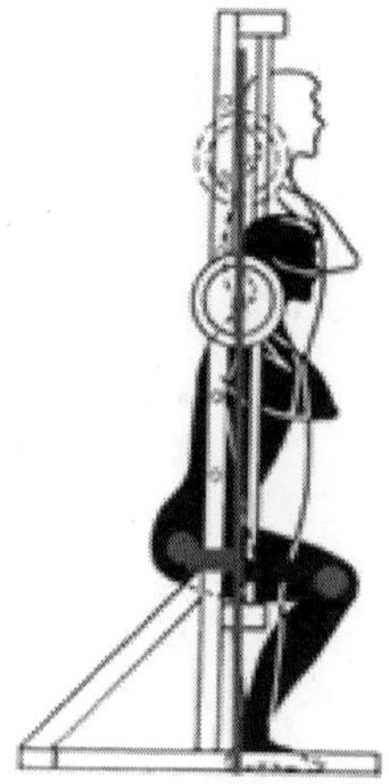

Fig movement pattern

1. You can see here the moment arm around the knees is substantially longer than that around the hip. This, over time, will encourage the quadriceps to become stronger and the gluteals to develop only a little. Unfortunately it will teach the person to squat with their thighs and not use their gluteals in a normal way. It will eventually bias the movement pattern and over time could cause knee injury or low back problems. You can also see the line of force is on the heels and not the mid-foot. This is not a natural position.
2. Although the torso is in good posture here, you usually see a flat back in the clubs. As the gluteals are not very active the core isn't always active. Also, the bar is stopping the person from moving forward as they can hang on to it, so there is little requirement for the trunk extensors to be active. This posture and it's its mechanics should not be encouraged.
3. Very few stabilisers will be used as the exercise has eliminated most natural sagittal plane requirements due to the fixed bar and virtually all front and transverse plane requirements.
4. If a person were to fail on this exercise it would be around the lumbar for stability. The most likely failure will be a very poor spinal position as you can keep lifting longer into muscular fatigue when you don't have to stabilise your position.

Interestingly when we ask why this squat is prescribed we often get told it is to help someone build up to free weight squatting because they can't yet squat, or it's to strengthen the gluteals and teach them the movement. None of this seems reasonable when you understand the biomechanics. The best way to help clients build up to free weight squatting is through teaching bodyweight squats with a range of movement that suits each individual client.

When you do this squat, the further you put your feet out the more the knee extensors (quadriceps) become active and the less the hip extensors (gluteals) do. As this occurs a force vector that creates knee shearing develops and increases such that if used over time the knee joint is almost guaranteed to get injured. You'll see this in a more pronounced fashion in the swiss ball squat.

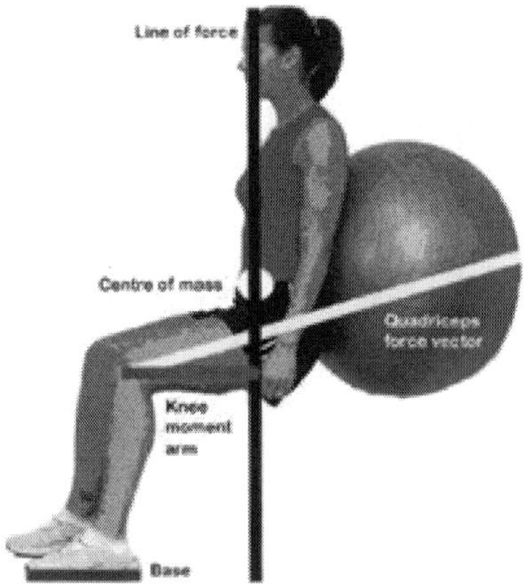

1. The line of force is marked going down through the centre of mass in blue. The base of support is a long way from the line of force and the centre of mass will be around the very top of the thighs. The knee moment arm is in green – basically this exercise is all quadriceps, and very little gluteals (if any) given there is virtually no moment arm for them to act around.

2. The yellow line represents the force vector of the quadriceps – we know a muscle shortens from it's its origin to it's its attachment. Because of the moment arm of the quadriceps and because the person must constantly push backward to keep the ball on the wall, there is constant shearing force at the knee joint. In order to cope with the quadriceps activity the inner unit of the knee will be constantly challenged (the inner unit includes the cruciate ligaments, lateral and collateral ligaments, and hamstrings). The torso is dormant as it is resting on the ball (if anything only the erector spinae may be active to hold the torso on the ball), and the gluteals are dormant as they have no way of contributing to the exercise. None of the learning, the brain is doing in this exercise is of any use, nor is it safe over the long term. So this is not a posture that we would ever encourage.

3. Although on a ball the transverse and frontal plane stability required is less than in a free standing squat. Sagittal plane stability is created through friction with the ball.

4. Failure in this exercise would likely be muscular and around the knee or hip. At the knee there may be significant discomfort causing the client to stop. At the hip the client may lose alignment (medially rotate the femur) resulting in the knees turning in.

Again this exercise is often prescribed as a regression from free body squatting. It seems, given the biomechanics, it is not.

Finally, given we like to 'progress' our clients by loading them, you can often see this exercise being completed with dumbbells in hand increasing the load acting through the line of force. It could be suggested that getting a baseball bat and smashing your client in the knee caps would be a more effective and upfront way of doing the damage people seem intent on causing with this exercise.

6.3 Torques in Equilibrium

A torque is an influence which tends to change the rotational motion of an object. One way to quantify a torque is

Torque = Force applied x lever arm

The lever arm is defined as the perpendicular distance from the axis of rotation to the line of action of the force.

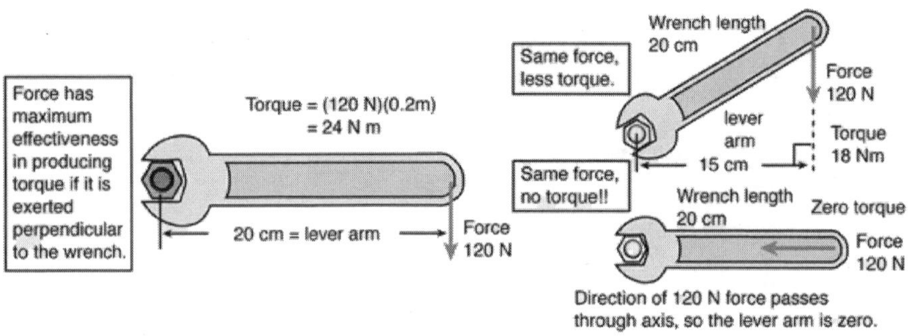

Fig Example of Torque acting on wrench

6.3.1 Conditions for Equilibrium

An object at equilibrium has no net influences to cause it to move, either in translation (linear motion) or rotation. The basic conditions for equilibrium are:

1. Net force = 0

x and y components of force may be separately set = 0.

$$\sum_i F_i = 0$$

Forces left = forces right
Forces up = forces down.

2. Net torque = 0

The axis may be chosen for advantage to eliminate some unknown forces..

$$\sum_i \tau_i = 0$$

The sum of the clockwise torques is equal to the sum of the counterclockwise torques.

Fig Conditions for Equilibrium

The conditions for equilibrium are basic to the design of any load-bearing structure such as a bridge or a building since such structures must be able to maintain equilibrium under load. They are also important for the study of machines, since one must first establish equilibrium and then apply extra force or torque to produce the desired movement of the machine. The conditions of equilibrium are used to analyze the "simple machines" which are the building blocks for more complex machines.

Why do long jumpers and high jumpers lower their centers of gravity before takeoff? What mechanical factors enable a wheelchair to remain stationary on a graded ramp or a sumo wrestler to resist the attack of his opponent? A body's mechanical stability is based on its resistance to both linear and angular motion.

6.3.1 Moment Arm

A moment arm (MA) determines the degree of effectiveness or influence of a force to produce or prevent the rotation of an object around an axis. The moment arm is the *shortest distance* from the *line of force* to the *axis*. Geometrically it is the perpendicular distance from the line of force to the axis, i.e. it is always *perpendicular* to the line of force.

To identify the length of the moment arm at any specific position during the movement of an object around an axis:

1. Identify the specific line of force in question (A)
   - ❖ Note: There are often multiple forces acting around a given axis
2. Identify the specific joint axis in question (B)
   - ❖ Note: There are often multiple axes affected by a line of force
3. Identify the line that is perpendicular to the line of force AND intersects the axis (C)
   - ❖ If the line of force (cable, gravity, muscle, etc.) does not appear to offer a perpendicular relationship to the axis at any point along it's depicted or

envisioned length, the line must be extended (mentally or graphically) in order to identify the moment arm (D)

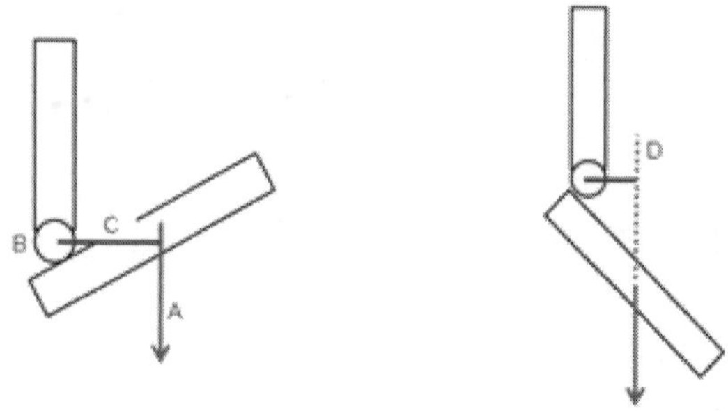

Fig Moment arm

Moment Arm and Torque

Torque is often described as "a twisting force" or "a force that causes rotation", but these descriptions are incomplete because a force can only cause rotation if it is applied at some distance from the axis, i.e. the line of force must not intersect the axis; there must be a *moment arm. "A force applied at some distance from an axis, causing or prevents a twist or rotation"* would be more accurate.

Torque is the product of the force applied around an axis times its distance from the axis or moment arm. (T = F x MA). Therefore, torque is directly proportional to both the magnitude or amount of force and its moment.

Moment Arm, Torque, and Moment

Moment is commonly considered synonymous with moment arm. However, colloquially, moment is also used as a synonym for torque. In essence the word moment represents the rotational influence of any specified force, e.g. *moment of inertia* is another term for *rotational inertia* which is represents how inertia is influenced by the mass's distance from the axis ($I_{r} = mr^{2}$)

Importance in Exercise

Every human movement is created by one or more joints rotating around their respective axes. Gross linear movements such as a squat (linear in that the center of mass must move vertically in a virtual straight line maintained over the base of support) are produced via multiple joints rotating in the required angular proportions. With this understanding it becomes necessary to recognize that virtually every force applied to or created within the human body is creating *torque*, and therein its influence upon either moving or stabilizing is *determined as much by the moment arm* of the force as the amount of force. Moment arm is one of the many reasons why the number on the weight does not accurately represent the true resistance and one of the reasons why what's moving and what's challenged can be unrelated.

Resistance Distribution

Regardless of the amount of load, the actual degree of challenge at each joint is determined by the moment to that joint. A squat is often called a total body exercise, but it does not challenge each part of the body equally (nor should it, as each joint and its associated musculature have drastically varying mechanical abilities). Although extremely oversimplified, these examples represent the distribution of the resistance between the hip and the knee. Although there are many factors involved (some that may be manipulated and some that may not) three very different experiences and outcomes can be expected from the squats as shown.

In squat A, the load is distributed equally between the hip and knee because the moments are equal. However, one is likely to experience greater knee extensor challenge than hip extensor challenge because the torque producing capabilities at the two joints are not equal, i.e. typically hip extensor strength is greater than knee extensor strength. In squat B load position has influenced body position in such a manner that a larger moment of resistance is created at the knee, which is often why it is stated in lay terms that *"front squats emphasize quads".*

In squat C, a larger moment is created at the hips (and lumbar spine). It is likely that this individual will experience squats as primarily a *"butt and low back exercise"* because that is exactly what is being challenged due to the moments of resistance. Fatigue/failure in the low back is likely to preclude any noticeable challenge to the quadriceps.

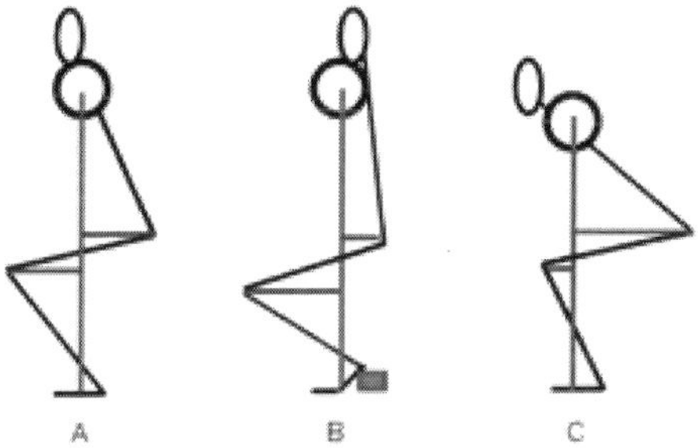

The examples of resistance distribution via comparative moment lengths are not intended as a complete analysis of squats, squat variations, related joint forces, exact muscular recruitment, etc. nor does this section or website offer recommendation or opinion regarding any version of squatting for any specific individual, goal, stage of progression, etc.

Strength Profiles

Strength is influenced by many factors, but at any given joint the fundamental mechanical elements of strength are the factors that influence internal torque production throughout a given range of motion: the *Length-Tension Relationship* of the involved musculature (the combined influences of active tension and passive tension as they vary throughout the shortening-lengthening cycle, as well as fatigue) and the associated *moments* of each muscle group (or the *resultant* thereof) as they change at every point in the range.

Ultimately, strength varies from start to finish in all human movements due in large part to the normal variations in the *moments of the internal force producers* that occur predictably and measurably throughout any considerable range of motion.

Resistance Profiles

As resistance to human movement is always *torque*, we *cannot* measure it solely by the *amount of force* (i.e. one's bodyweight, the number on the end of the dumbbell, or attached to the cable). The changing moment of resistance to every involved joint (be it moving or static) creates a predictable and measurable variation in challenge throughout the range of motion. If the source of resistance utilized is one that changes in magnitude throughout the range (tubing, mass/inertia, etc.) then this further influences the torque of resistance.

Strategic manipulation of the moment of resistance (in concert with a strategic choice of the source of resistance based upon its properties) is vital to creating optimal and appropriate Resistance Profiles for specific individuals and goals. This is the key to creating "full range challenge" without which "full range of motion" of minimal value.

Examples

Cams

As it rotates a cam manipulates either the moment arm of *resistance* (below) or *effort*. If designed appropriately it will generate a strategically appropriate alteration in (or maintenance of) torque throughout each point of the range of motion.

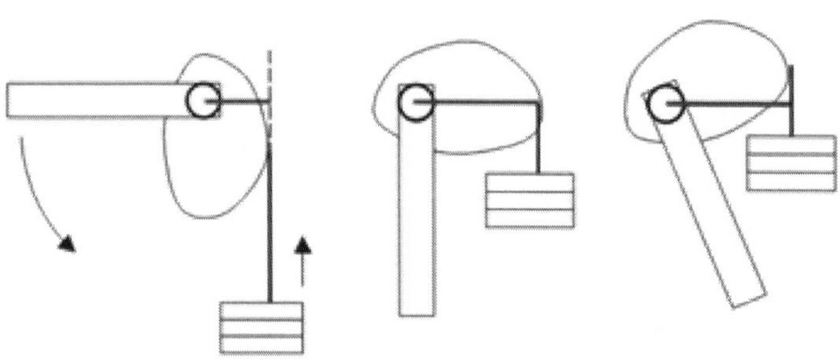

Fig moment arm of resistance

Now the final output of the machine is the sum of the combined changes occurring within the machine throughout the chosen range, which may include multiple cams and/or lever systems. It is vital to recognize that the machine's final output does not constitute the resistance profile of the exercise, only the *magnitude profile* of the resistance. The actual resistance profile can only be determined by comparing these magnitude changes to the concurrent changes in *moment* to the user's joints throughout the range. This is where the force from the device actually becomes torque to the body and therein, the resistance for the exercise.

Internal Torque Production

The patella is one of the best examples of both moment arm and structural influence in one's "strength". Often referred to as an anatomical pulley, the patella really acts more like an *anatomical cam*, dramatically altering the moment of the quadriceps collective line of force. Surgical removal of the patella (for decades the common treatment for fracture) rendered the knee extensors virtually powerless to straighten the knee under load regardless of the tension producing capabilities of the quadriceps.

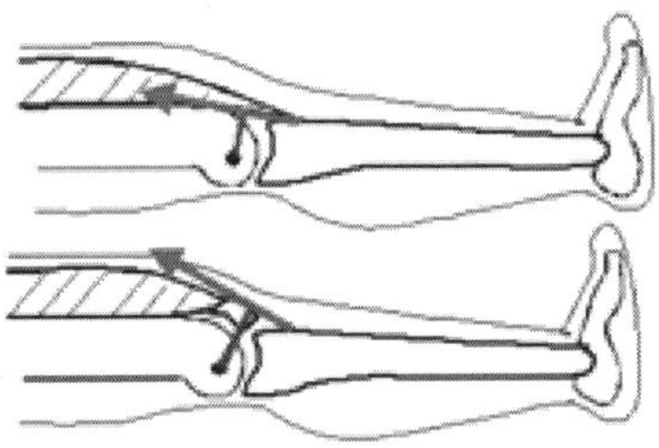

Characteristics of Moments

1. If a direction of force remains consistent while a limb rotates at a joint, the *moment arm will change with every degree of motion*. This simply means that the force will have different degrees of influence at every position comprising the range of motion.

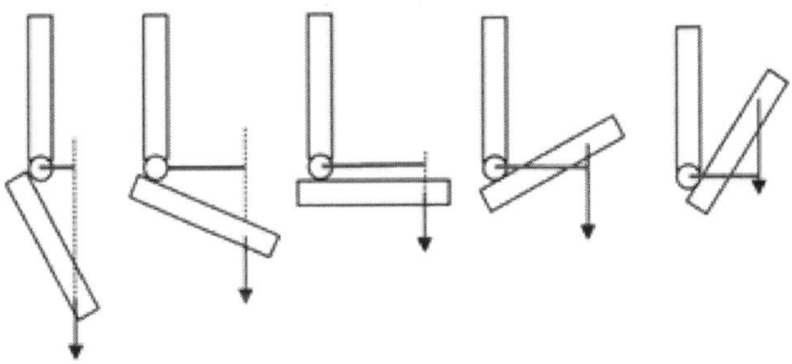

2. There will be a different moment for *each force* that acts upon a joint axis: internal force or external force, effort force or resistance force.

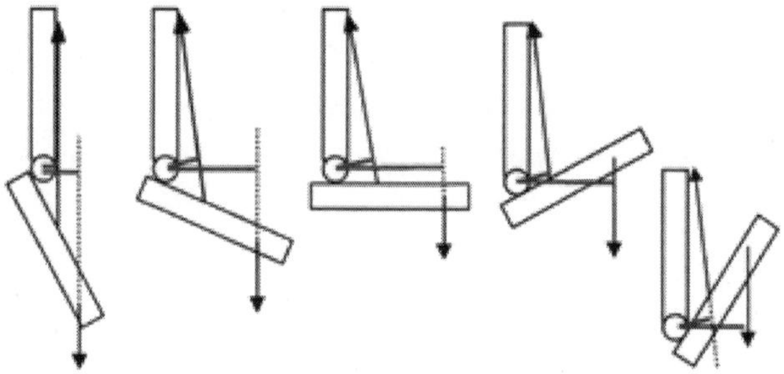

3. A single force can influence every involved joint. There will be a separate moment to each joint, the length of which will determine the degree of challenge to the musculature surrounding that joint.
4. A moment arm can never be longer than its associated lever arm. I.e. the length of the lever will be the maximal length of the moment. It can be shorter than the lever but never longer.
5. The maximal moment for a given scenario will occur when the force angle is "optimal" (90°). Any degree less than or greater than 90 will produce a smaller moment.

   A 45° force angle and a 135° force angle will produce the same moment because they are the same distance from 90°.
6. When the force angle is 90° the lever arm and the moment arm coincide, i.e. are the same length and occupy the same space.

Moment arm is the most vital mechanical factor that is consistently ignored by the exercise industry, experts and consumers alike. When presented in formal study it is with such poor examples and lack of reverence that one must assume that the professors themselves don't really understand its importance in exercise. Below are just a few of the numerous examples of moment arm neglect or misunderstanding.

In case of balanced see-saw, moment arm (perpendicular distance from the line of force to the axis) happens to coincide with the lever arm (the distance from the point of application of the force to the axis) because the force angle is 90°. If the seesaw is shown unbalanced it becomes obvious that these two are very different.

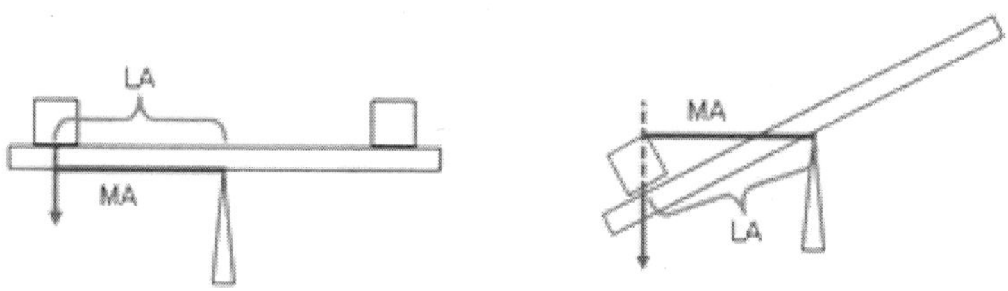

Free Weights Are Not Constant Resistance

As seen, *free weight uses constant load based on barbells or dumbbells throughout range of motion.* The only free weight that is constant resistance is one that is not moving. A weight that is moving will be a variable resistance due to the potentially dramatic influence of inertial effects, and a weight moving around an axis will always be a variable resistance due to the constantly changing moment arms to each involved joint.

The most obvious characteristic of elastic resistance is that the more the elastic component is stretched, the greater will be the resistance. It is true that the greater the stretch the greater the tension or magnitude of resistance, but this is only half the equation because we must remember that resistance to the body is torque, not just force alone. The moment is an equally influential factor and if the moment to a particular joint gets close enough to zero, the resistance at that joint will be insignificant. In the examples below the greatest length of tubing will produce virtually no resistance at the shoulder.

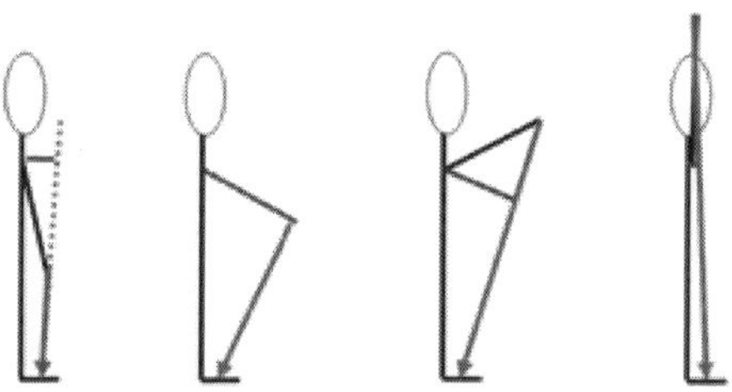

Fig greatest length of tubing

Chest Press is a standing, forward pressing, upper extremity exercise with tubing wrapped around the torso just below the arm pit. It shows tubing wrapped around the torso produces a zero moment at the shoulder joint throughout the entire range, and therefore no resistance in that plane at the shoulder joint which would require the chest (pectoralis major) to be recruited any more than would horizontally adducting the shoulder in a standing position without the tubing (a position in which the movement is unresisted by gravity). There is a significant moment resisting elbow extension, at least at the start.

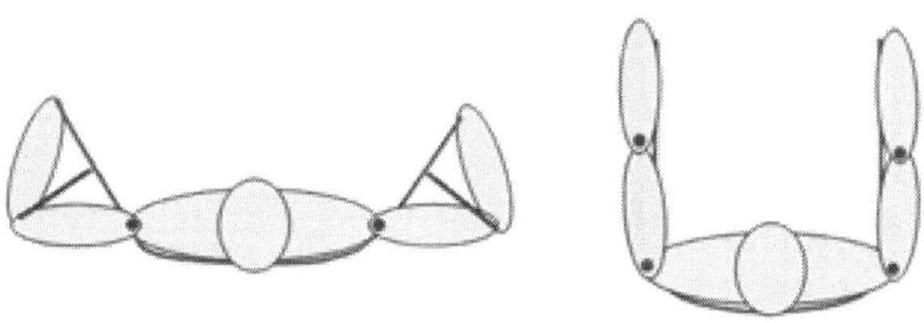

Fig tubing wrapped around the torso

6.4 Center of Gravity

6.4.1 Gravity

Gravity is a very important force to consider when dealing with biomechanics. It is constantly affecting the body in both static and dynamic movement. It can be looked at as the gravitational downward pull on an object. Whereas the center of mass is the mean position of matter in a body or system. Gravity, like all forces, has a point of application, a magnitude, and a direction.

The point of application is the center of gravity or center of mass (COG or COM). This changes relative to how many segments are involved. For instance, the COG of the forearm alone will be different then the COG of the arm, forearms, and hand. The magnitude is dependent on ~~the~~ the orientation of the segment in space and is proportional to the mass of the involved segment. The direction of gravity is commonly referred the line of gravity (LOG). This direction is always vertically downward towards the center of the earth, regardless of the orientation of the segment/object to space.

Gravity has a profound effect on biomechanics. The gravitational force acts on the levers of the body to create torque at various body segments and joints. Thus, it encompasses almost all basic biomechanical concepts when acting on the human body. We use this knowledge of gravity and anatomy to manipulate muscle positions for manually grading muscle strength, movement analysis, and specific therapeutic interventions.

6.4.2 Center of gravity and Center of mass

Center of gravity is the point through which the force of gravity acts on an object or system. In most mechanics problems the gravitational field is assumed to be uniform. The center of gravity is then in exactly the same position as the center of mass. The terms center of gravity and center of mass tend to often be used interchangeably since they are often at the same location.

Center of mass

Center of mass is a position defined relative to an object or system of objects. It is the average position of all the parts of the system, weighted according to their masses. For simple rigid objects with uniform density, the center of mass is located at the centroid. For example, the center of mass of a uniform disc shape would be at its center. Sometimes the center of mass doesn't fall anywhere on the object.

6.4.3 Centre of gravity in human body

Centre of gravity (COG) of the human body is a hypothetical point around which the force of gravity appears to act. It is point at which the combined mass of the body appears to be concentrated. Because it is a hypothetical point, the COG need not lie within the physical bounds of an object or person. One subjective way (there are objective measures) to approximate the COG of an object is to visualize it balancing on one finger.

In the anatomical position, the COG lies approximately anterior to the second sacral vertebra. However, since human beings do not remain fixed in the anatomical position, the precise location of the COG changes constantly with every new position of the body and limbs. The bodily proportions of the individual will also affect the location of the COG.

The direction of the force of gravity through the body is downward, towards the centre of the earth and through the COG. This line of gravity is important to understand and visualise when determining a person's ability to successfully maintain Balance. When the line of gravity falls outside the Base of support (BOS), then a reaction is needed in order to stay balanced.

When the line of gravity is within the BOS, an object or person is said to be stable. When the line of gravity falls outside the BOS, the object or person is said to be unstable. Given that the line of gravity must fall within the BOS in order to satisfy the criteria for stability, the following factors should be considered:

- A larger BOS increases stability (the line of gravity must move a greater distance to fall outside the BOS)
- A lower COG increases stability (it's unlikely that the line of gravity will fall outside the BOS)

Center of Gravity in the anatomical position is just anterior to the second sacral vertebrae. However, it's important to note that COG/COM is conceptual. It is not a piece of anatomy; it is constantly changing with motion. With every movement and change of position the COG changes the way joints react and muscles perform. For example, the picture below shows the how the COG changes when the hip is flexed to 90°. This brings the COG anteriorly and creates a clockwise torque at the hip joint. To counteract this moment of torque created by the COG, the contralateral hip extensors must fire to keep the body statically erect.

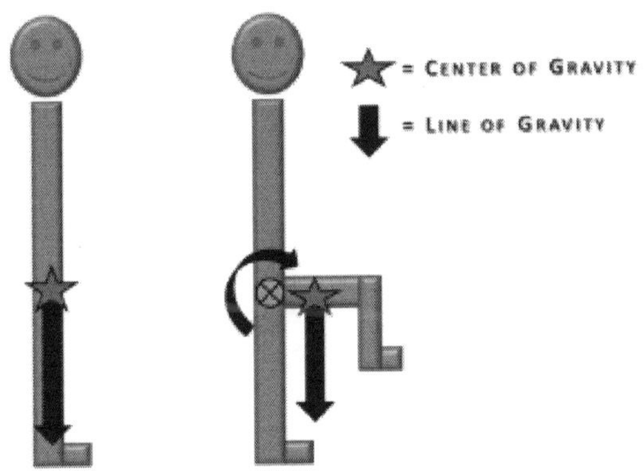

Fig Change in COG and LOG with hip flexion

6.4.4 Center of Gravity outside the Body

Another consideration is that COG does not have to be a part of the body. Many exercises and movements cause the body's COG to become displaced outside of the body. This would create an increase in the lever arm of the resistance force. Proper lifting mechanics are simply a means of decreasing this lever arm as much as possible. As seen the figure below, bending forward at the hip causes the COG to actually fall outside of the human body.

Fig stability of level

Applying basic biomechanics to gravity will give a better understanding of how postures, movement, and exercises affect the body. A simple change in position of one extremity can cause a significant change in joint forces, muscle activation patterns, and neuromuscular coordination demands. Keep this in mind when assessing movements and implementing interventions.

- Gravity is the force that attracts a body toward the earth (downward)
- Center of Gravity/Center of Mass is the point of application of the gravitational force
- The direction of gravity is the Line of Gravity (LOG) and is always perpendicular to the ground
- COG in the anatomical position is just anterior to the second sacral vertebrae
- COG is not a piece of anatomy, it is conceptual and changes with different positions and motion
- COG is relevant to the body segments that are included in the system
- COG may be located outside of the body
- The LOG position relevant to the BOS is the basis for stability

6.4.5 Stability and Equilibrium

Equilibrium are of three types:

Stable equilibrium

- occurs when an object is placed in such a position that any disturbance effort would raise its COM
- tend to fall back its original position
- e.g. BOS or ¯COM

unstable equilibrium

- occurs when an object is placed in such a position that any disturbance effort would lower its COM
- tend to fall into a more stable position

neutral equilibrium

- occurs when an object is placed in such a position that any disturbance effort would not change the level of its COM
- tend to fall into a more stable position

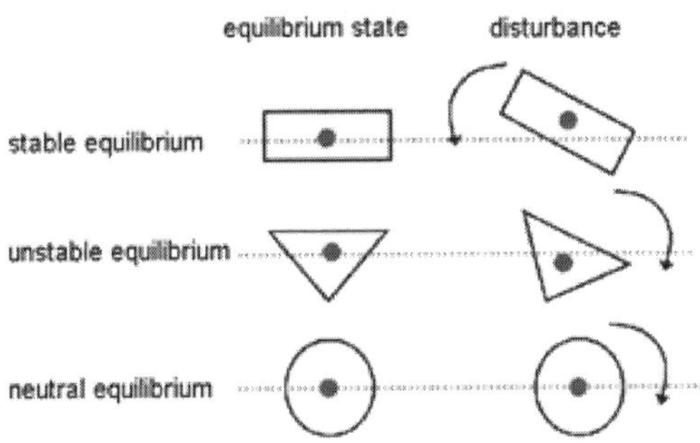

Factors Affecting Stability

Size and shape of base of support (BOS)

- wide-base stance
- tandem stance: standing with one foot ahead the other
- stance with crutches

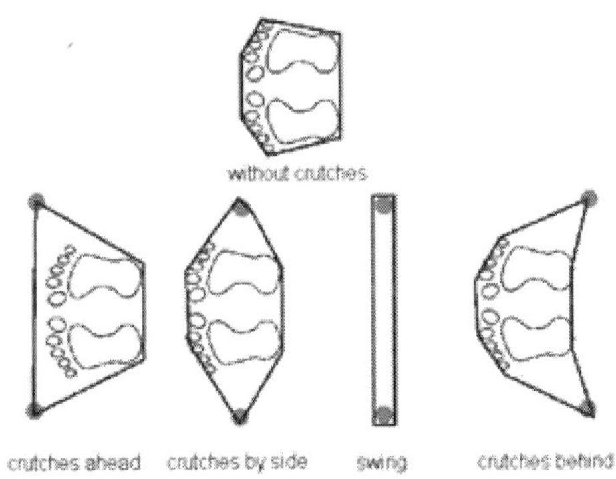

Effect of Crutches Use on Base of Support

- height of COM
- relationship of COG to BOS
- mass of body
- friction
- segmental alignment
- sensory input
    - visual
    - vestibular system
    - proprioception
    - other somatosensory system
- psychological or mental status
- muscle activities
- postural muscle: the muscle that acts to prevent collapse of the skeleton
    - slow twitch
    - fatigue resistant
- phasic muscle : fast muscle
    - physiological and pathological factors

6.4.6 Qualitative Analysis

A natural application of angular kinetics and anthropometrics is the determination of the center of gravity of the body. The center of gravity is the location in space where the weight (gravitational force) of an object can be considered to act. The center of small rigid objects (pencil, pen, bat) can be easily found by trying to balance the object on your finger. The point where the object balances is in fact the center of gravity, which is the theoretical point in space where you could replace the weight of the whole object with one downward force. There is no requirement for this location to be in a high-mass area, or even within or on the object itself. Think about where the center of gravity of a basketball would be. The center of gravity of the human body can move around, because joints allow the masses of body segments to move. In the anatomical position, the typical location of a body's center of gravity in the sagittal plane is at a point equivalent to 57

Spine and Low-Back Pain

One of the most common complaints is low-back pain. The medical literature would say that the etiology (origin) of these problems is most often idiopatic (of unknown origin). The diagnostic accuracy of advanced imaging techniques like magnetic resonance imaging (MRI) for identifying spinal abnormalities (e.g., disk herniation) that correlate with function and symptoms of low-back pain is poor. The causes of low-back pain are complicated and elusive. Biomechanics can contribute clues that may help solve this mystery. Mechanically, the spine is like a stack of blocks separated by small cushions.

Stability of the spine is primarily a function of the ligaments and muscles, which act like the guy wires that stabilize a tower or the mast of a boat. These muscles are short and long and often must simultaneously stabilize and move the spine. Total spine motion is a summation of the small motions at each intervertebral level. Biomechanical studies of animal and cadaver spines usually examine loading and rotation between two spinal levels in what is called a motion segment. Individuals even exhibit different strategies for rotation of motion segments in simple trunk flexion movements, so that neuromuscular control likely plays an important role in injury and rehabilitation.

As reported, x-ray measurements of the buckling of single spinal segment that occurred during a heavy deadlift, biomechanics research uses computer models and EMG to understand how muscles and loads affect the spine, and the nature of this motion segment buckling. The information is combined with occupational, epidemiological, neurologic, and rehabilitative research to understand the development and treatment of low-back pain.

There are two main methods used to calculate the center of gravity of the human body, and both methods employ the equations of static equilibrium. One lab method, which requires a person to hold a certain body position, is called the reaction change or reaction board method. The other method used in research is called the segmental method. The segmental method uses anthropometric data and mathematically breaks up the body into segments to calculate the center of gravity.

The reaction board method requires a rigid board with special feet and a scale (2D) or scales (3D) to measure the ground reaction force under the feet of the board.

The "feet" of a reaction board are knife-like edges or small points similar to the point of a nail. A 2D reaction board, a free-body diagram, and static equilibrium equations to calculate the center of gravity in the sagittal plane are illustrated in Figure.

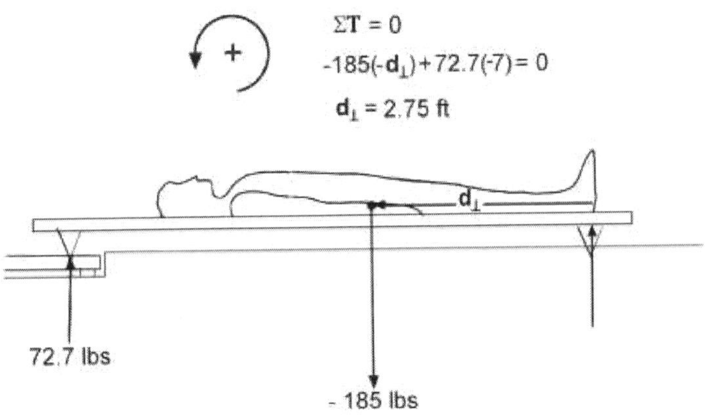

Fig free-body diagram to calculate center of gravity

Note that the weight force of the board itself is not included. This force can be easily added to the computation, but an efficient biomechanist zeros the scale with the board in place to exclude extra terms from the calculations. The subject in above figure weighs 185 pounds, the distance between the edges is 7 feet, and the scale reading is 72.7 pounds. With only three forces acting on this system and everything known but the location of the center of gravity, it is rather simple to apply the static equilibrium equation for torque and solve for the center of gravity (d±). Note how the sign of the torque created by the subject's body is negative according to convention, so a negative d± (to the left) of the reaction board edge fits this standard, and horizontal displacement to the left is negative. In this case, the subject's center of gravity is 2.75 feet up from the edge of the reaction board. If the subject were 5.8 feet in height, his center of gravity in this position would be 47% of his or her height.

In the segmental method, the body is mathematically broken up into segments. The weight of each segment is then estimated from mean anthropometric data. For example, weight of the forearm and hand is 2.52 and 2.07% for a man and a woman, respectively. Mean anthropometric data are also used to locate the segmental centers of gravity (percentages of segment length) from either the proximal or distal point of the segment.

Segment Centers of Gravity (X,Y)
Shank/feet (3.2,6.6)
Thighs (5.0,10.5)
Head/Arms/Trunk (11.0,9.0)

$$\Sigma T_o = 0$$

$$182\,(d_\perp) - 22(3.2) - 36(5) - 124(11) = 0$$

$$d_\perp = 8.9 \text{ in}$$

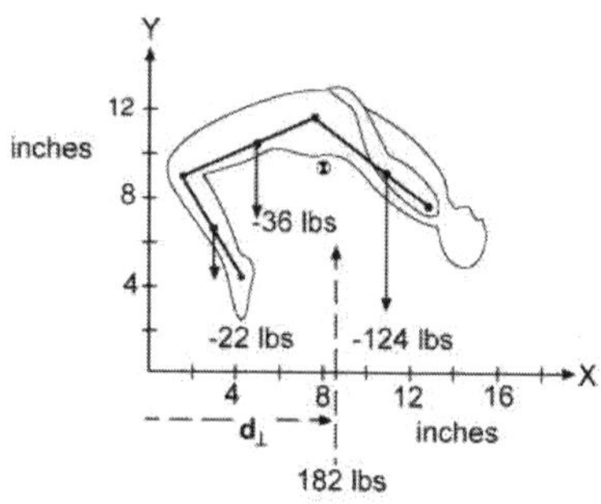

Fig Calculating the horizontal position of the whole body center of gravity of a high jumper

Figure shows the calculation of the center of gravity of a high jumper clearing the bar using a three-segment biomechanical model. This simple model (head+arms+trunk, thighs, legs+feet) illustrates the segmental method of calculating the center of gravity of a linked biomechanical system. Points on the feet, knee, hip, and shoulder are located and combined with anthropometric data to calculate the positions of the centers of gravity of the various segments of the model. Most biomechanical studies use rigid-body models with more segments to more accurately calculate the whole-body center of gravity and other biomechanical variables. If a biomechanist were studying a high jump with high-speed video (120 Hz), a center of gravity calculation much like this would be made for every image (video snapshot) of the movement.

The segmental method is also based on static equilibrium. The size and location (moment arm) of the segmental forces are used to calculate and sum the torques created by each segment. If this body posture in the snapshot were to be balanced by a torque in the opposite direction (product of the whole bodyweight acting in the opposite direction times the center of gravity location: 182 • d±), the total torque would be zero. By applying the law of statics and summing torques about the origin of our frame of reference, we calculate that the person's bodyweight acts 8.9 inches from the origin. These distances are small because the numbers represent measurements on an image. In a 2D biomechanical analysis, the image-size measurements are scaled to real-life size by careful set-up procedures and imaging a control object of known dimensions.

Finding the height of the center of gravity is identical, except that the y coordinates of the segmental centers of gravity are used as the moment arms. Students can then imagine the segment weight forces acting to the left, and the height of the center of gravity is the y coordinate that, multiplied by the whole bodyweight acting to the right, would cancel out the segmental torques toward the left. Based on the subject's body position and the weights of the three segments, guess the height in centimeters of the center of gravity. Did the center of gravity pass over the bar? Finish the calculation in above figure to check your guess. The segmental method can be applied using any number of segments, and in all three dimensions during 3D kinematic analysis. There are errors associated with the segmental method, and more complex calculations are done in situations where errors (e.g., trunk flexion/extension, abdominal obesity) are likely.

Chapter 7
Angular Kinematics

7.1 Angular Position and Displacement

7.1.1 Angular Quantities

Consider a disk contained in the xy plane and rotating counterclockwise about a fixed axis passing through its center. A point B in the disk at a distance r from the center will move along the dash circle of radius r, figure a).

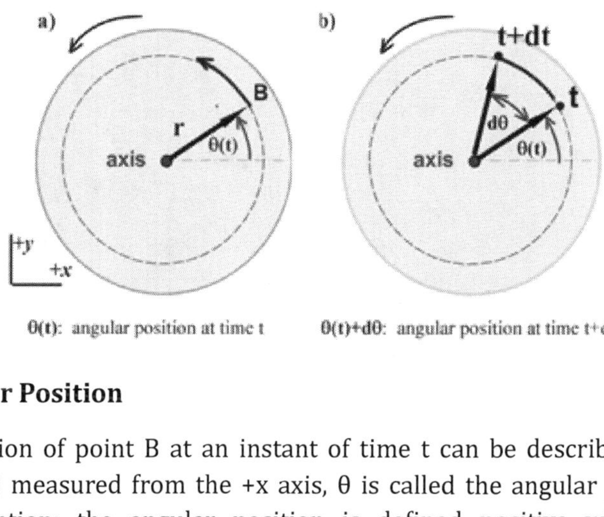

θ(t): angular position at time t θ(t)+dθ: angular position at time t+dt

7.1.2 Angular Position

The position of point B at an instant of time t can be described in terms of the angle θ(t) measured from the +x axis, θ is called the angular position of the point. Convention: the angular position is defined positive when measured counterclockwise with respect to the +x axis.

7.2 Angular Velocity

At a later time t + dt, the point B has moved along the circle and its angular position has changed by dθ and is now θ(t+dt) = θ(t) + dθ, figure b).

$$\omega = \frac{d\theta}{dt}$$

For a given time interval dt, the faster the rotation of the object larger is the change in angular position. The angular velocity is defined as the rate of change of the angular position and its noted with the letter omega:

7.2.1 Direction of the angular velocity

The angular velocity is a vector. Its direction is parallel to the axis of rotation, therefore the angular velocity vector is perpendicular to the plane where the circle described by point B is contained.

In the figures below, the rotation of a disk is described from the point of view of an observer looking from the +z axis. As shown in the figure:

- For counterclockwise rotation: θ(t) increases with time, then $\omega > 0$ and points towards the +z axis - towards the observer.
- For clockwise rotation: θ(t) decreases with time, then $\omega < 0$ and points towards the -z axis - away from the observer.

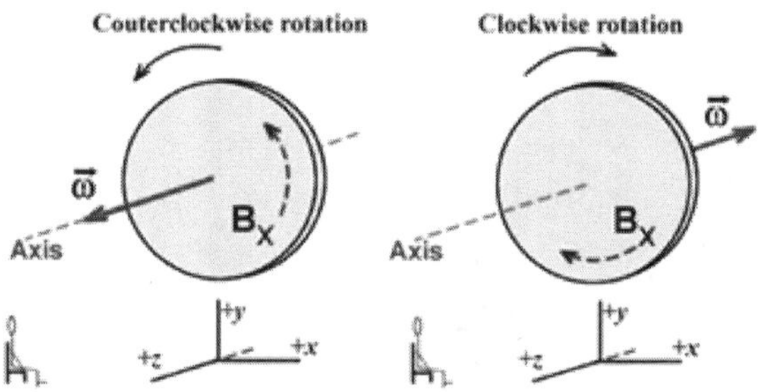

Note that what it looks rotating counterclockwise from one side is rotating clockwise from the other side. To obtain the correct direction of the angular velocity vector we will use the right hand rule.

Use the right hand rule where direction of the angular velocity vector can be obtained by curling the fingers of your right hand along the direction of the rotation of object and your thumb will point to the direction of the angular velocity vector.

7.3 Angular Acceleration

If the angular velocity changes with time then there is an angular acceleration noted by the greek letter α and defined as:

$$\alpha = \frac{d\omega}{dt} = \frac{d^2\theta}{dt^2}$$

7.3.1 Direction of the angular acceleration vector

The angular acceleration is a vector and its direction is either parallel or anti-parallel to the angular velocity vector. In analogy to one dimensional motion, if the angular acceleration vector is in the same direction as the angular velocity vector then the rotation is speeding up. If the vectors are in opposite direction then the rotation is slowing down.

The angular velocity and the angular acceleration are the quantities that define the state of rotation of the rigid body. **ALL** the points inside the rigid body have the same angular velocity and angular acceleration. However, the linear speed and the acceleration of a given point in the rigid body depends on the position of the point relative to the axis of rotation. Consider again the point B in the disk. The faster the disk rotates faster the point will move along the circle.

Between the instant **t** and **t+dt** the rigid object has rotate an angle **dθ**. At the same time, point B has moved along the arc length **ds** which is related to the change in angular position **dθ** by:

$$ds = r d\theta$$

The further point **B** is from the axis or rotation the larger is **r** and larger is the arc length ds that the point must covered in the time **dt**.

7.3.2 Linear or tangential velocity

The linear or tangential velocity is the rate of change of the position vector of point B:

$$\vec{v} = \frac{d\vec{r}}{dt}$$

The term linear is used to distinguish this quantity from the angular velocity which is the rate of change of the angular position.

The direction of the linear velocity is always tangent to the circle described by point B. For this reason it is also called tangential velocity. From now on we will use the term 'velocity' or 'speed' of a point when we refer to the linear or tangential velocity or speed but we will mention the term 'angular' when we talk about $\omega = d\theta/dt$.

The magnitude of the velocity of point B can be obtained by taking the derivative with respect t time of ds:

$$v = \frac{ds}{dt} = r\frac{d\theta}{dt} = r\omega$$

Note that the speed of point B is proportional to the angular speed and the distance from the center. The further is the point from the axis of rotation, larger is its speed. A point further from the axis must change its angular position in the same time interval as a point close to the axis of rotation, therefore it has to cover a larger arc length in the same time interval.

7.4 Linear Acceleration

7.4.1 Linear or tangential acceleration

The acceleration vector of point B is the rate of change of the velocity vector. It has two components, a radial component, ar, that points towards the center of the circle, and a component tangent to the circle, a_{tan}:

$$\vec{a} = \frac{d\vec{v}}{dt} = a_r\hat{r} + a_{tan}\hat{\theta}$$

The **radial component** is due to the **change in the direction** of the velocity vector and is given by:

$$a_r = -\frac{v^2}{r}$$

and, the **tangential component** is due to the **change of the magnitude** of the velocity vector:

In general the acceleration of a point moving in a circle is given by:

7.4.2 Uniform circular motion" α = 0 and ω = constant

In the particular case of uniform circular motion when ω is constant then α = 0 and atan = 0. The acceleration vector has only a radial component:

$$\vec{a} = -\frac{v^2}{r}\hat{r}$$

7.5 Rotational Inertia

Inertia is the bodies body's resistance to change in motion and is related to mass. If you want to get an object with a large mass moving you have to apply more force than if you want to get a tiny object moving. It really is that simple. Likewise if you want to slow down or stop a heavy object you will need to apply a greater force to it than if you want to slow down a small object.

For rotation, there is a similar concept called rotational inertia. Rotational inertia is the resistance of an object to being rotationally accelerated. Rotational inertia is dependent not only upon mass, but also upon the distance of mass from the center of rotation.

- A measure of the distribution of mass about the axis of rotation.
- If the mass is far from the axis, the Moment of Inertia is large.
- If the mass is close to the axis, the Moment of Inertia is small.

A certain object or system of objects has some rotational inertia. The rotational inertia about a certain axis is called the moment of inertia. Because the distance from the mass to the axis of rotation is important, a single object can have very different moments of inertia depending upon the axis about which it rotates. The moment of inertia for an object is defined as:

$$I = \sum m_i * r_i^2$$

Where i is the number of objects.

In this, r is the distance from the axis of rotation to the mass. As can be seen in the equation, the moment of inertia is dependent upon the mass of the object and the square of the distance from the mass to the axis of rotation. Just like how linear kinematics has equations of motion, rotational kinematics has analogous equations of motion. For example, Newton's second law for linear motion is:

$$\vec{F} = m * \vec{a}$$

A similar rotational equation takes the form:

$$\vec{\tau} = I * \vec{\alpha}$$

where is the torque, is the moment of inertia, and is the angular acceleration. Here, the moment of inertia is the analog of the mass term in Newton's second law. Similarly, the moment of inertia is present in the other important equations of rotational motion:

$$Angular\ momentum = \vec{L} = I * \vec{\omega}$$

$$Kinetic\ energy = \frac{1}{2} I * \omega^2$$

where is the angular velocity of the object.

7.6 Angular Momentum

Angular can be defined as the movement of a mass when it is rotating or spinning. You may have seen a situation when a person in a tucked position spins faster or than someone in an extended position.

To explain the movement of a mass when it is rotating, we must first understand angular momentum, this is the movement of a mass around an object and is calculated by multiplying angular velocity with the moment of inertia.

Angular momentum= angular velocity x moment of inertia.

Angular velocity is referred to as the speed of rotation, in other words, how quickly something rotates. The moment of inertia is referred to as the resistance of a body to change state when rotating, in other words its reluctance to rotate. If a body has a high reluctance to rotate (high moment of inertia), it would have a lower angular velocity. If a mass has a low resistance to rotate, then it will have a higher angular velocity.

The moment of inertia can be calculated by multiplying the mass of a product x distance from the axis.

Moment of inertia= mass x distance from the axis of rotation.

A body is unlikely to lose mass during a movement but it can change its distance or distribution from the axis of rotation. If the mass moves closer to the axis of rotation then the moment of inertia decreases, resulting in the angular velocity increasing. If a mass moves further away from the axis of rotation, then the moment of inertia increases and angular momentum decreases.

7.6.1 Features

Angular momentum is a characteristic of a rotating body about a certain axis and is dependent upon the moment of inertia about that axis and the angular velocity about that axis.

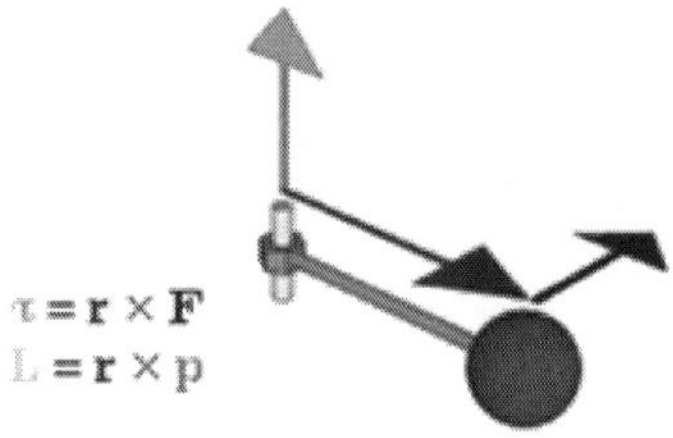

Force $\Sigma F = ma$ and Torque $\Sigma\tau = I\alpha$

- Newton's Laws also apply to angular motion.
- For every linear term, there is an equivalent angular term.
- For example, torque is the angular effect of force. Just like a net force produces an acceleration resisted by the mass, a net torque produces an angular acceleration resisted by the moment of inertia.

Linear Impulse $\Sigma F\Delta t = m\Delta v$ and Angular Impulse $\Sigma\tau\Delta t = \Delta I\omega$

- A net force acting for a period of time produces a linear impulse that results in a change in linear momentum.
- Likewise, a net torque acting for a period of time produces an angular impulse that results in a change in angular momentum.
- Where angular momentum is the product of the moment of inertia and angular velocity

Here $\Delta I\omega$ is the change in angular momentum produced by the resultant torque ($\Sigma\tau$) acting for Δt.

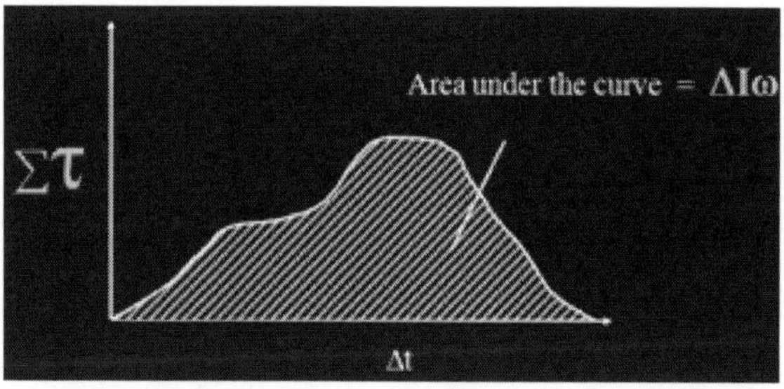

Fig $\Sigma\tau\Delta t = \Delta I\omega$

7.6.2 Conservation of Angular Momentum

- Just as linear momentum is conserved when no external forces act on a system, angular momentum is conserved when no external torques act on a system.
- Understanding the above statement is unimaginably crucial to your success.
- This would be the case when a system is freely rotating on a turntable, or travelling through the air.

7.6.3 Generating Angular Momentum with an External Torque

Generating the greatest amount of angular momentum prior to leaving the ground is important in such events as figure skating.

To maximize the amount of angular momentum generated prior to take-off, the athlete should maximize their moment of inertia about the axis they are trying to rotate about. This would be the longitudinal axis in figure skating.

Fig skating

The increase in (I) means that for any given torque the athlete's body will take a longer time to rotate which means a longer time for the torque to act which leads to a greater change in angular momentum

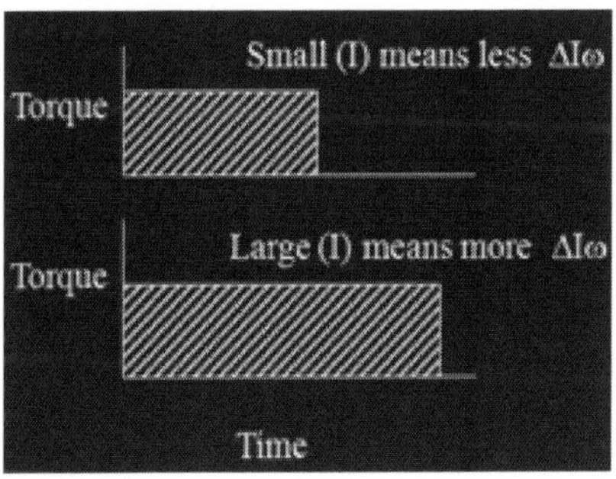

7.6.4 Sports

Angular momentum is the product of inertia (a body's resistance to changes in motion) and that body's angular velocity. As a result, in order to maintain a constant angular momentum, the figure skater can manipulate their body in order to increase its inertia, while subsequently decreasing its angular velocity or visca vice versa. In the figure skating example, the figure skaters will utilize their arms in order to decrease their inertia (bring the arms closer to their body) which in turn causes them to increase their rate of spin or angular velocity. As they near the end of the spin, they will move their arms further away from their body (increasing their inertia) which in turn will cause a decrease in their angular velocity and slow down their rate of spin.

This principle also plays a vital role in the success of elite level divers. When the diver moves from a layout position to a tuck position, they are ultimately manipulating their moment of inertia in order to adjust their angular velocity. If the divers timing is off, they will not make a clean entrance into the pool. This is part of the reason why you will see divers break form (ie. flex the knees) near the completion of their jump in order to increase their velocity in hopes of making a cleaner entrance into the pool. Elite level divers are extremely gifted in knowing where they should be when coming out of a dive in order to make a clean entrance into the water. This is where repetition and quality feedback from the coach and/ or Biomechanist can help to improve the diver's performance and overall score on a dive.

Chapter 8
Anatomical Concepts

8.1 Material Characterization of Tissues

Regenerative medicine aims to engineer materials to replace or restore damaged or diseased organs. The mechanical properties of such materials should mimic the human tissues they are aiming to replace; to provide the required anatomical shape, the materials must be able to sustain the mechanical forces they will experience when implanted at the defect site. Although the mechanical properties of tissue-engineered scaffolds are of great importance, many human tissues that undergo restoration with engineered materials have not been fully biomechanically characterized. Several compressive and tensile protocols are reported for evaluating materials, but with large variability it is difficult to compare results between studies. Further complicating the studies is the often destructive nature of mechanical testing. Whilst an understanding of tissue failure is important, it is also important to have knowledge of the elastic and viscoelastic properties under more physiological loading conditions.

Patients are increasingly waiting for various organ transplantations to treat failing or injured organs. However, with the shortage of suitable donor organs, regenerative medicine is aiming to create alternative solutions for patients with end-stage organ failure. Regenerative medicine aims to meet this clinical need by engineering materials to act as tissue substitutes, including soft tissues, such as cartilage and skin. To create a successful material to restore damaged tissues, the replacement material should mimic the properties of the native tissue it is going to replace. Once surgically implanted, the material will need to provide anatomical shape to the tissue defect and thus, the mechanical properties of the material are vital. For example, a material replacing auricular cartilage should have the appropriate mechanical properties to prevent compression by the overlying skin2. Similarly, a material to replace nasal cartilage will need to have adequate mechanical properties to prevent collapsing during breathing. However, despite the importance of mechanical properties when manufacturing materials for implantation, little evidence has focused on characterizing the mechanical properties of different human tissues.

Mechanical testing regimes can be used to establish the compressive, tensile, bending, or shear properties of a tissue. Skin is a highly anisotropic, viscoelastic, and nearly incompressible material. Commonly excised skin is tested using uniaxial tensile methodologies, where a suitably shaped strip of skin is gripped at both ends and stretched while the load and extension are recorded.

Since the major component of all soft tissues is interstitial water, the mechanical response of cartilage is strongly related to the flow of fluid through the tissue. Soft tissues such as cartilage have been traditionally tested using compression testing. The methods for testing in compression are quite varied, with confined, unconfined, and indentation being the most prevalent as shown in figure. Within confined compression, a cartilage sample is placed in an impervious, fluid-filled well and loaded through a porous plate. Since the well is non-porous, flow though the cartilage is in the vertical direction. In unconfined compression, the cartilage is loaded using a non-porous plate onto a non-porous chamber, forcing the fluid flow to be predominantly radial. Indentation is the most frequently used method for evaluating the biomechanical properties of cartilage. It consists of an indenter, smaller than the surface of the specimen being tested, that is brought down onto the specimen. Indentation has many advantages over other methods of compression, including the fact that indentation can be performed in situ, enabling the test to be more physiological as shown.

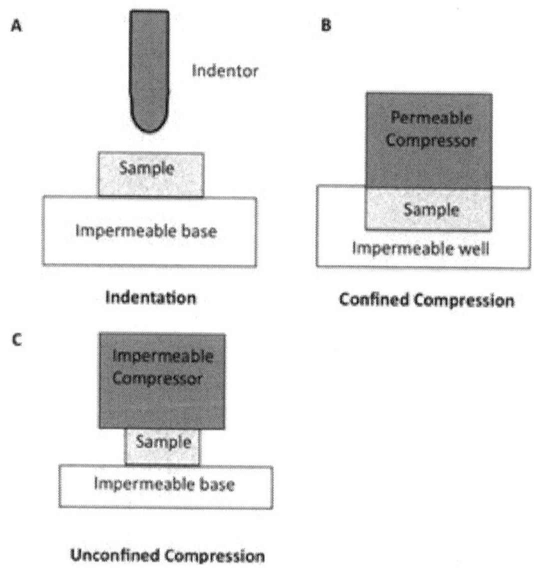

Fig different compression methodologies

To understand the compressive and tensile properties of a tissue, the Young's elastic modulus is typically calculated by analyzing the linear portion of the stress-strain curve, indicating the elastic resistance to compression or tension, irrespective of specimen size. Both tensile and compressive testing regimes can vary according to the load or deformation applied and the rate of both such parameters. At present, there are many different testing protocols to assess tissue mechanics, which makes it extremely difficult to interpret or compare results from different studies. Furthermore, many mechanical methods currently focus on characterizing the mechanical properties of the tissue by testing the specimen to destruction.

Skin is a complex, multi-layered material, divided into three main layers: the epidermis, dermis, and hypodermis. The mechanical properties of skin tissue have recently been evaluated using in vivo assessments. However, protocols of tensile testing can be utilized to understand the skin biomechanics of excised skin. Such tests can provide information to model stress-strain relationships, since the boundary conditions can be defined. Typically, in vitro testing regimes use high strains to characterize the material to failure, whereas in vivo systems use low strain ranges4. When comparing biomechanical values for excised skin in tension, there is a large variability between different studies, ranging from 2.9-150 MPa. Large differences between subjects are expected due to natural biological variation, but differences in protocol regimes can also compound these natural biological differences. For example, differences in loading rates between protocols will cause variation, as greater loading rates cause less time for the fluid to flow out, resulting in a higher stiffness. The preparation, excision, and handling protocols of the skin tissue will also cause differences in the mechanical properties. This protocol demonstrated for testing skin provides an alternative method for researchers to characterize skin tissue. It provides a few advantages, including the ability to identify the elastic and viscoelastic properties of skin tissue in one mechanical test, allowing for a greater understanding of the skin in a short amount of time. Furthermore, the same test can be applied to tissue-engineered replacements to manufacture constructs with similar biomechanical properties as native skin.

Indentation testing provides an attractive option compared to confined compression testing for understanding the biomechanics of cartilage. Indentation has the ability to preserve the physiological structure of the cartilage and thus provides values that mimic those of a clinical setting. Using indentation, it is also possible to test the cartilage while still attached to the underlying bone. Indentation also allows for physiological testing of cartilage as in vivo. When two cartilage surfaces approach each other, the edges surrounding the area of contact "bulge" due to water under the contact area being displaced laterally after compressive deformation occurs.

Cartilage indentation must be conducted with an indenter with a smaller radius than the cartilage sample to allow for similar bulging. The size of the indenter should also be at least times the sample size to ensure that the cartilage reacts as if it were part of an indefinite sample. Using an indenter much smaller than the radius of the sample diameter eliminates any edge effects present in specimen creation. In addition, indentation avoids possible experimental errors caused by testing cartilage defects damaged by sample extraction. Indentation also does not involve deep sample preparation, such as confined compression, allowing small, thin pieces of cartilage to be tested. Furthermore, the non-destructive method of indentation means that it has a potential application in the clinical setting as a diagnostic tool after validation and verification studies have been performed.

8.2 Mechanics of Skeletal Muscles

8.2.1 Muscles

Muscle is the only tissue capable of actively developing tension. This characteristic enables skeletal, or striated, muscle to perform the important functions of maintaining upright body posture, moving the body limbs, and absorbing shock. Because muscle can only perform these functions when appropriately stimulated, the human nervous system and the muscular system are often referred to collectively as the neuromuscular system.

The four behavioral properties of muscle tissue are extensibility, elasticity, irritability, and the ability to develop tension. These properties are common to all muscle, including the cardiac, smooth, and skeletal muscle of human beings, as well as the muscles of other mammals, reptiles, amphibians, birds, and insects.

8.2.2 Properties

The properties of extensibility and elasticity are common to many biological tissues. As shown in figure, extensibility is the ability to be stretched or to increase in length, and elasticity is the ability to return to normal length after a stretch. Muscle's elasticity returns it to normal resting length following a stretch and provides for the smooth transmission of tension from muscle to bone.

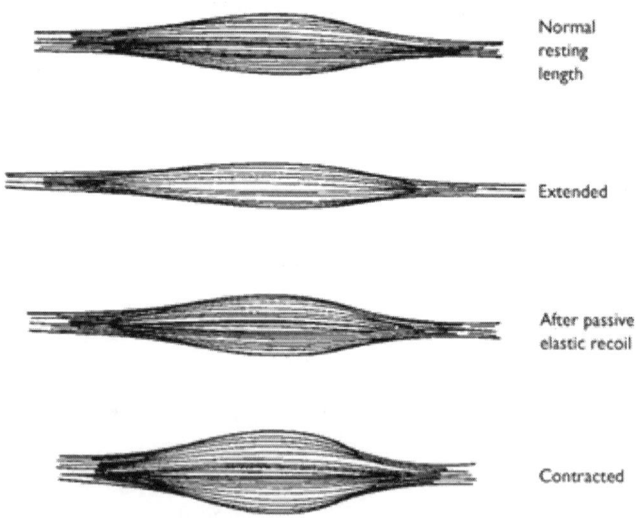

Fig characteristic properties of muscle tissue

8.2.3 Features

Muscles are like a series of Russian dolls, with each structure containing a smaller structure. Additionally, each structure is wrapped around with its own, dedicated collagen layer, called the extracellular matrix. Each muscle is enveloped by the epimysium, and is built up of many muscle fascicles. Muscle fascicles are themselves wrapped around by the perimysium and are comprised of many muscle fibers, which do not always run the length of the muscle fascicle that contains them. Muscle fibers (which are single cells) are enveloped by the endomysium, and can reach 400mm long and 0.1mm wide. They contain myofibrils suspended and supported by a cytoskeleton. The myofibrils are covered by the sarcoplasmic reticulum and are around 1µm wide. They are made up of chains of sarcomeres in series, and run the full length of the individual muscle fibers.

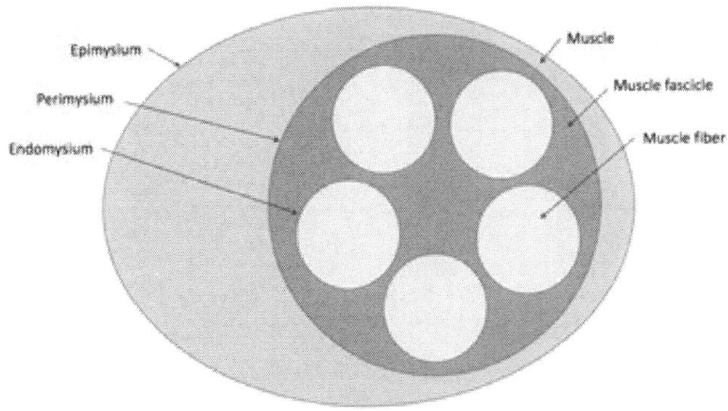

Fig Structure

Muscle cell biology can be initially confusing, as most cells in the body are small and contain only one nucleus. In contrast, muscle cells are very large and contain many thousands of nuclei, as well as the long chains of giant molecules that make up the myofibrils.

Whole Muscles

Whole muscles are classically represented in a fusiform model, in which a collection of fibers run together in parallel from the origin tendon, to the insertion tendon. In this way, it is thought that the fibers shorten and pull the insertion closer to the origin. In fact, most muscles are not structured like this at all. They are generally not fusiform, but are rather either uni-pennate or multi-pennate. In such structures, the muscle fibers run at angles to the main line of action, from one side of the muscle to the other. These structures are described by two separate sets of characteristics: muscle architecture, and the regions of the muscles.

Muscle architecture is the organization of the muscle fibers within the muscle relative to the line of force generation. It comprises three main aspects: normalized fiber length, pennation angle, and physiological cross-sectional area. Muscle architecture is the single most important factor when it comes to determining a muscle's function, force production capacity and contraction velocity, particularly in comparison with other muscles.

Most muscles are not comprised of single compartments in which all the muscle fibers run from one end to the other. Rather, each muscle is made up of several segments, which display different features. For example, as reported, gluteus medius has four separate regions (anterior, anterior-middle, posterior-middle, and posterior). These compartments have different nerve branches and varying pennation angles (from +33 degrees in the anterior region to -30 degrees in the posterior region), which means that they best suited to slightly different tasks.

Muscle fiber type can be measured using immunohistochemistry to assess the type of myosin present in individual muscle fibers. Some isoforms of myosin display a faster contraction velocity than others, and thereby myosin heavy chain (MHC) type determines the force–velocity relationship of the sarcomeres in a myofibril. Some MHC types have faster contraction speeds than others, and these are termed type II muscle fibers. Athletes having a higher proportion of type II muscle fibers are thought to have an advantage at strength and power sports.

8.2.4 Features of Skeletal muscle

Skeletal muscle in normal, healthy individuals contains around 6% collagen by weight. The collagen content of muscles exists at various different levels, embedded as fibers within the extra-cellular matrix at each level. The whole muscle is covered by a thick layer of connective tissue called the epimysium, while the muscle fascicles within the muscle are covered by collagen fibers called the perimysium, and each individual muscle fiber is surrounded by its basement membrane, which is also variously called an endomysium, sarcolemma, or plasmalemma. Each of these layers of connective tissue are made up of collagen fibers held within a matrix of hydrated proteoglycans.

In many muscles at rest, the collagen fibers of the epimysium and perimysium are oriented at angles of approximately 55 degrees? To the long axis of the muscle fibers. However, in some (especially pennate muscles), the collagen fibers in the epimysium (the surface layer) run parallel to the long axis of the muscle, which behaves in a similar way to a tendon. The endomysium surrounding individual muscle fibers deforms nonlinearly with increasing sarcomere length such that it does not transmit force between muscle fibers very effectively at resting muscle lengths, and transmits force between muscle fibers at long lengths by means of shear forces.

Unlike the epimysium and endomysium, muscles can differ in respect of how much perimysium they have, and how it is distributed. The perimysium can be very thin or very thick and can surround either small or large groups of muscle fibers, which leads to differences in perimysial collagen content in the same muscles from between 0.43% – 4.6% of dry weight. This variation is much greater than that of the endomysium and has important consequences for muscle material properties. Additionally, the collagen and proteoglycan content of the perimysium is different from that of the epimysium and endomysium. The perimysium contains primarily type I collagen and decorin (the same as tendons), while the epimysium and endomysium contain both types I and III collagen, and a range of proteoglycans. Again, the higher amount of type I collagen in the perimysium may be contributory to different functions, as it provides greater friction and could therefore contribute more to shear loading. It also appears that the connection points between the individual muscle fibers and the perimysium are co-localized with focal adhesions. Focal adhesions are mechanoreceptors that may have a role in hypertrophic cellular signaling. Together, these features indicate that the perimysium likely plays a role in the transmission and detecting of muscle force.

Although each muscle fiber has an endomysium, which surrounds it and provides structure and containment for the myofibrils within it, it also has a set of structures that link the individual myofibrils together inside the endomysium. This set of structures is called the cytoskeleton. The cytoskeleton is made up of parts that are found outside the sarcomere (the exosarcomeric cytoskeleton) and parts that are found intertwining within the sarcomeres themselves (the endosarcomeric cytoskeleton).

The exosarcomeric cytoskeleton is made up of various proteoglycans, including desmin, vimentin, and synemin. It can be thought of as a set of scaffolding that sits at right angles to the myofibrils, and links the contracting chains of sarcomeres to the sheath that is the endomysium. The chains of sarcomeres are made up of thin (actin) and thick (myosin) myofilaments. Each sarcomere joins to the next one in the series at a junction called a Z disk. The actin myofilaments are connected to this Z disk by crosslinking to α-actinin, and the Z disk is attached to the extracellular matrix through focal adhesions, which are also called costameres.

The endosarcomeric cytoskeleton is predominantly made of two giant proteins, titin and nebulin. Both titin and nebulin run parallel to and intertwine with the actin and myosin myofilaments, providing structural support points, and attachment points for various molecules. They are thought to function primarily to prevent the chain of sarcomeres from overextending itself and therefore breaking, and they likely also have key roles in passive force generation.

The giant protein nebulin was originally thought to function primarily as a molecular ruler that monitored the length of the thin (actin) myofilament. In support of this idea, several studies have reported that the lengths of thin (actin) myofilaments tend to be similar to those of nebulin within the same muscle, and mice genetically modified without nebulin do not survive because they experience muscular dystrophy. More recently, some reports have provided contrary evidence, showing that the length of nebulin does not necessarily always correspond to the full length of the thin (actin) myofilament, and changing the length of nebulin does not necessarily alter the length of the thin (actin) myofilament. Additionally, nebulin seems to have other roles. It appears to be involved in the regulation of calcium flux through the sarcoplasmic reticulum, may augment the interaction of actin and myosin within the sarcomere, and seems to be important for the structure of the Z disk structure, which is the boundary between adjacent sarcomeres in series. Moreover, as expected, nebulin seems to be elastic and therefore likely provides passive support for the thin (actin) myofilament, which likely increases force production on the descending arm of the length-tension curve and also protects the sarcomeres from being disrupted.

8.3 Continuum Mechanics Concepts in Modeling of large deformation

Basic concepts of continuum mechanics have been established more than five decades ago, the 21st century faces many new and exciting potential applications of continuum mechanics that go way beyond the standard classical theory. Continuum mechanics is a combination of mathematics and physical laws that approximate the large-scale behavior of matter that is subjected to mechanical loading. It is a generalization of Newtonian particle dynamics, and starts with the same physical assumptions inherent to Newtonian mechanics; and adds further assumptions that describe the structure of matter.

8.3.1 Newtonian reference frame

In classical continuum mechanics, the world is idealized as a three dimensional Euclidean space (a vector space consisting of all triads of real numbers (x1,x2,x3). A *point* in space is identified by a unique set of three real numbers. A Euclidean space is endowed with a *metric,* which defines the distance between points: d=√xixi. Vectors can be expressed as components in a ***basis*** - {**e**1,**e**2,**e**3} of mutually perpendicular unit vectors. Physical quantities such as force, velocity, acceleration are expressed as vectors in this space. A *Cartesian Coordinate Frame* is a fixed point O together with a basis. A Newtonian reference frame is a particular choice of Cartesian coordinate frame in which Newton's laws of motion hold.

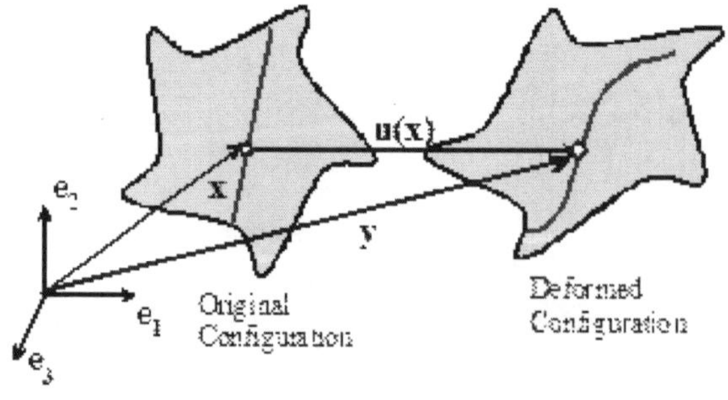

Modeling an object as a continuum assumes that the substance of the object completely fills the space it occupies. Modeling objects in this way ignores the fact that matter is made of atoms, and so is not continuous; however, on length scales much greater than that of inter-atomic distances, such models are highly accurate. Fundamental physical laws such as the conservation of mass, the conservation of momentum, and the conservation of energy may be applied to such models to derive differential equations describing the behavior of such objects, and some information about the particular material studied is added through constitutive relations.

Continuum mechanics deals with physical properties of solids and fluids which are independent of any particular coordinate system in which they are observed. These physical properties are then represented by tensors, which are mathematical objects that have the required property of being independent of coordinate system. These tensors can be expressed in coordinate systems for computational convenience.

8.3.2 The Continuum

Materials, such as solids, liquids and gases, are composed of molecules separated by space. On a microscopic scale, materials have cracks and discontinuities. However, certain physical phenomena can be modeled assuming the materials exist as a continuum, meaning the matter in the body is continuously distributed and fills the entire region of space it occupies. A continuum is a body that can be continually sub-divided into infinitesimal elements with properties being those of the bulk material.

The validity of the continuum assumption may be verified by a theoretical analysis, in which either some clear periodicity is identified or statistical homogeneity and ergodicity of the microstructure exists. More specifically, the continuum hypothesis/assumption hinges on the concepts of a representative elementary volume and separation of scales based on the Hill–Mandel condition. This condition provides a link between an experimentalist's and a theoretician's viewpoint on constitutive equations (linear and nonlinear elastic/inelastic or coupled fields) as well as a way of spatial and statistical averaging of the microstructure.

When the separation of scales does not hold, or when one wants to establish a continuum of a finer resolution than that of the representative volume element (RVE) size, one employs a statistical volume element (SVE), which, in turn, leads to random continuum fields. The latter then provide a micromechanics basis for stochastic finite elements (SFE). The levels of SVE and RVE link continuum mechanics to statistical mechanics. The RVE may be assessed only in a limited way via experimental testing: when the constitutive response becomes spatially homogeneous.

Matter is idealized as a continuum, which has two properties: (i) it is infinitely divisible (you can subdivide some region of the solid as many times as you wish); and

(ii) it is locally homogeneous – in other words if you subdivide it sufficiently many times, all sub-divisions have identical properties (eg mass density). A continuum can be thought of as an infinite set of vanishingly small particles, connected together.

Both the existence of a Newtonian reference frame, and the concept of a continuum, are mathematical idealizations. Experimental evidence suggest that the laws of motion based on these assumptions accurately approximate the behavior of most solid and fluid materials at length scales of order mm-km or so in engineering applications. In some cases continuum models can also approximate behavior at much shorter length scales (for volumes of material containing a few 1000 atoms), but models at these length scales often require different relations between internal forces deformation measures in the solid to those used to model larger volumes.

8.4 Deformation and Strain

8.4.1 Introduction

The stress equilibrium equations, although derived for the deformed state of that material, did not entail any assumptions about the material or the type of deformation it incurs. We learned that for most situations, the stress equilibrium equations are indeterminate, and we can't solve for material stresses directly from these equations. We therefore need to consider other aspects of the material and tissue mechanics. The next logical step is to examine then how the material deforms. It is at this point that we introduce our first significant assumptions concerning continuum mechanics. If we make no assumptions about the size of deformation, then the resulting strain tensor is valid for every situation. However, this strain tensor is nonlinear and leads to complex analysis. Therefore, if the deformation is small (typically less than 3-4%), then we can use a small deformation analysis, which is linear and simpler to use. In tissue mechanics, hard tissues fit under the small deformation model, but most soft tissues typically undergo large deformation (strains > 5%).

8.4.2 Small Deformation Strain Tensor

We first consider the case of small deformation, since this is applicable to bone and is much less complex than large deformation. Consider the deformation of an infinitesimal cube in the x1 direction:

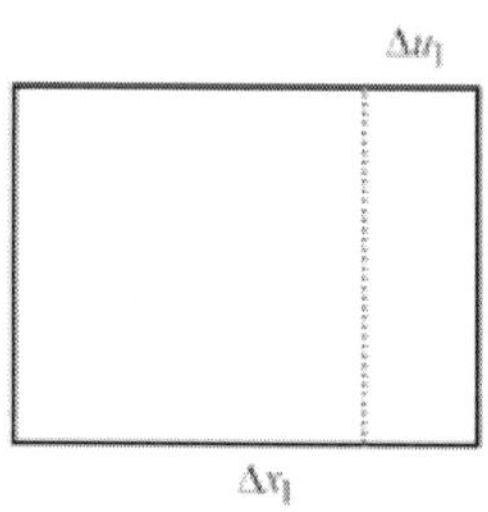

Fig deformation of an infinitesimal cube

We know that if the displacements are small, then we can approximate the normal strain in the direction as:

$$\varepsilon_{11} \approx \frac{\Delta u_1}{\Delta x_1}$$

If we in the limit allow both Δx1 and Δu1 both to approach zero, then we arrive at the following strain displacement relationship:

$$\varepsilon_{11} = \frac{\partial u_1}{\partial x_1}$$

This corresponds to our notion that strain is defined as the change in length over length. Note that this is an approximation of strain for small deformations. Thus, it is our definition of strain that renders this discussion for the small deformation part of small deformation elasticity. We can also do the same change in length for the x2 and x3 directions to arrive at:

$$\varepsilon_{22} = \frac{\partial u_2}{\partial x_2}$$

$$\varepsilon_{33} = \frac{\partial u_3}{\partial x_3}$$

Next, let us consider a pure shear deformation as shown as the infinitesimal cube below:

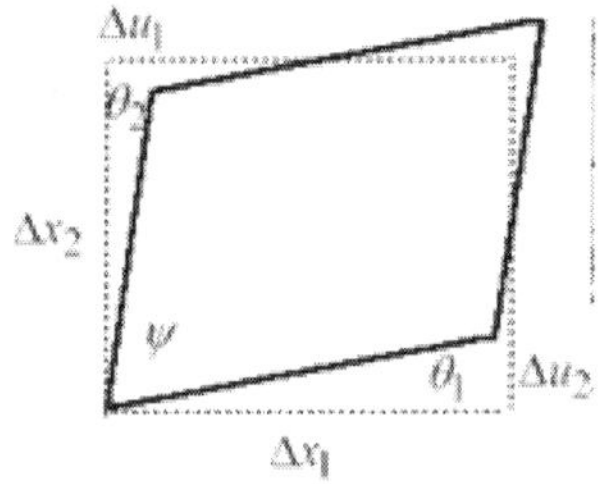

Fig shear deformation in cube

Here, the total deformation is the sum of the angles q1 and q2, and can be written as:

$$\gamma_{12} = \frac{\pi}{2} - \psi = \theta_1 + \theta_2$$

Note that q_1 is defined to equal q_2,$q_1 = q_2$. Now using trigonometry we see that the tangent of q_1 and the tangent of q_2 can be written as:

$$\tan\theta_1 = \frac{\Delta u_2}{\Delta x_1}$$

$$\tan\theta_2 = \frac{\Delta u_1}{\Delta x_2}$$

If we assume that the angle of deformation is small (again referring back that this section is *small deformation* strain), we know that the tangent of a small angle is equivalent to the angle itself, assuming we define the angle in radians. Thus, we may write the above as:

$$\tan\theta_1 \approx \theta_1 = \frac{\Delta u_2}{\Delta x_1} \; (\mathit{for\, small}\; \theta_1)$$

$$\tan\theta_2 \approx \theta_2 = \frac{\Delta u_1}{\Delta x_2} \; (\mathit{for\, small}\; \theta_2)$$

We can thus rewrite the angular deformation from above as:

$$\frac{1}{2}\gamma_{12} = \varepsilon_{12} = \frac{1}{2}(\theta_1 + \theta_2) \approx \frac{1}{2}\left(\frac{\Delta u_2}{\Delta x_1} + \frac{\Delta u_1}{\Delta x_2}\right)$$

Note that we have added an extra definition e12. This is the definition of tensorial shear strain that is one half g12 that is the definition of engineering shear strain. If we taken the limit as all delta quantities go to zero, then we have the following definition of the shear strain:

$$\frac{1}{2}\gamma_{12} = \varepsilon_{12} = \frac{1}{2}\left(\frac{\partial u_1}{\partial x_2} + \frac{\partial u_2}{\partial x_1}\right)$$

by definition shear strains are symmetric as we see below:

$$\varepsilon_{12} = \frac{1}{2}\left(\frac{\partial u_1}{\partial x_2} + \frac{\partial u_2}{\partial x_1}\right) \Rightarrow \varepsilon_{21} = \frac{1}{2}\left(\frac{\partial u_2}{\partial x_1} + \frac{\partial u_1}{\partial x_2}\right) \therefore \varepsilon_{12} = \varepsilon_{21}$$

The definition of the other shear strains follows directly as:

$$\varepsilon_{13} = \frac{1}{2}\left(\frac{\partial u_1}{\partial x_3} + \frac{\partial u_3}{\partial x_1}\right)$$

$$\varepsilon_{31} = \frac{1}{2}\left(\frac{\partial u_3}{\partial x_1} + \frac{\partial u_1}{\partial x_3}\right)$$

$$\varepsilon_{23} = \frac{1}{2}\left(\frac{\partial u_2}{\partial x_3} + \frac{\partial u_3}{\partial x_2}\right)$$

$$\varepsilon_{32} = \frac{1}{2}\left(\frac{\partial u_3}{\partial x_2} + \frac{\partial u_2}{\partial x_3}\right)$$

Given the definition of normal strains, and the shear strains given above, we can write a compact definition of small strain in indicial notation as:

$$\varepsilon_{ij} = \frac{1}{2}\left(\frac{\partial u_i}{\partial x_j} + \frac{\partial u_j}{\partial x_i}\right)$$

Note that the small deformation strain is a second order tensor just like the Cauchy stress tensor. It has nine components for 3D. We can write the small deformation strain tensor in matrix form as:

$$\begin{bmatrix} \varepsilon_{11} & \varepsilon_{12} & \varepsilon_{13} \\ \varepsilon_{21} & \varepsilon_{22} & \varepsilon_{23} \\ \varepsilon_{31} & \varepsilon_{32} & \varepsilon_{33} \end{bmatrix} = \begin{bmatrix} \frac{1}{2}\left(\frac{\partial u_1}{\partial x_1} + \frac{\partial u_1}{\partial x_1}\right) & \frac{1}{2}\left(\frac{\partial u_1}{\partial x_2} + \frac{\partial u_2}{\partial x_1}\right) & \frac{1}{2}\left(\frac{\partial u_1}{\partial x_3} + \frac{\partial u_3}{\partial x_1}\right) \\ \frac{1}{2}\left(\frac{\partial u_2}{\partial x_1} + \frac{\partial u_1}{\partial x_2}\right) & \frac{1}{2}\left(\frac{\partial u_2}{\partial x_2} + \frac{\partial u_2}{\partial x_2}\right) & \frac{1}{2}\left(\frac{\partial u_2}{\partial x_3} + \frac{\partial u_3}{\partial x_2}\right) \\ \frac{1}{2}\left(\frac{\partial u_3}{\partial x_1} + \frac{\partial u_1}{\partial x_3}\right) & \frac{1}{2}\left(\frac{\partial u_3}{\partial x_2} + \frac{\partial u_2}{\partial x_3}\right) & \frac{1}{2}\left(\frac{\partial u_3}{\partial x_3} + \frac{\partial u_3}{\partial x_3}\right) \end{bmatrix} =$$

$$\begin{bmatrix} \frac{\partial u_1}{\partial x_1} & \frac{1}{2}\left(\frac{\partial u_1}{\partial x_2} + \frac{\partial u_2}{\partial x_1}\right) & \frac{1}{2}\left(\frac{\partial u_1}{\partial x_3} + \frac{\partial u_3}{\partial x_1}\right) \\ \frac{1}{2}\left(\frac{\partial u_2}{\partial x_1} + \frac{\partial u_1}{\partial x_2}\right) & \frac{\partial u_2}{\partial x_2} & \frac{1}{2}\left(\frac{\partial u_2}{\partial x_3} + \frac{\partial u_3}{\partial x_2}\right) \\ \frac{1}{2}\left(\frac{\partial u_3}{\partial x_1} + \frac{\partial u_1}{\partial x_3}\right) & \frac{1}{2}\left(\frac{\partial u_3}{\partial x_2} + \frac{\partial u_2}{\partial x_3}\right) & \frac{\partial u_3}{\partial x_3} \end{bmatrix}$$

8.4.3 Reference (Initial) and Deformed Configurations

The first step in defining large deformation strain measures is to define the relationship between what is known as the reference, initial or undeformed configuration of a body, and the *deformed* configuration of the body. The reference or undeformed configuration is the condition of the body in 3D space before loads have been applied to it.

The deformed configuration is the location and shape of the body after loads have been applied to it. It is important to note that the body may undergo rigid body motion in addition to strain when loades are placed on it. An illustration of the relationship between the initial and deformed configuration is shown below:

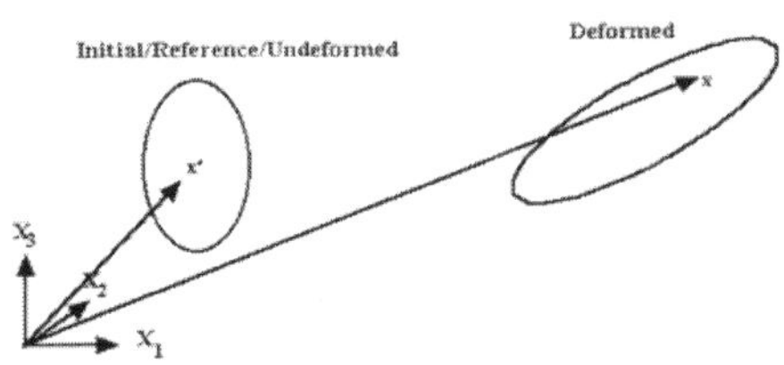

Fig Relationship between a point in the Reference and Deformed Configuration

Note that we have defined a vector in 3D space x' in the reference configuration of the body and a vector x in the deformed configuration of the body. We note that the relationship between the two position vectors in space, which represent the locations of a point in the reference configuration and a point in the deformed configuration, is the displacement vector, as shown below:

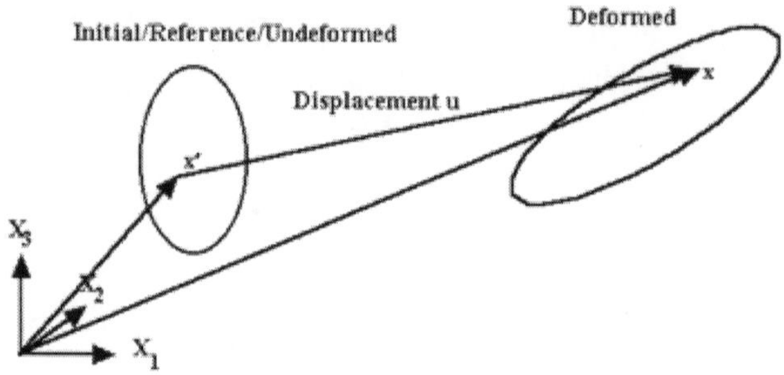

Fig vector addition

By vector addition, we can directly write the relationship between the position vectors in the initial and deformed configuration:

$$x_i = x_i^{'} + u_i$$

Deformation Gradient

The above results tells us how a point displaces from the reference to the deformed configuration. However, we would also like to know how a piece of material, again, an infinitesimal piece, is stretched and rotated as the body moves from the reference to the deformed configuration. The infinitesimal material vector in the reference configuration, dx', is shown in the reference configuration. The material vector after it has been stretched and rotated, dx, is shown in the deformed configuration below:

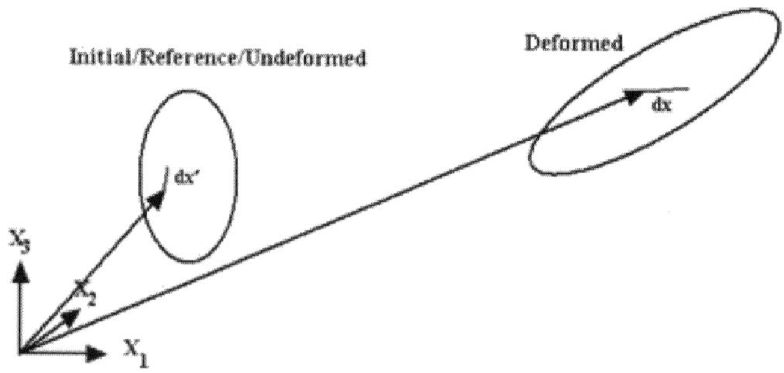

Fig infinitesimal material vector in the reference configuration

Since the length of dx' can change when going to the deformed configuration as well as its orientation, we can say that dx' deforms into dx. The question then becomes how to relate dx in the deformed configuration into dx' of the reference configuration. This can readily be done through the chain rule as:

$$dx_i = \frac{\partial x_i}{\partial x'_j} dx'_j$$

The above equation gives a relationship between a material vector in the undeformed and deformed configuration. We define the mapping itself as the deformation gradient tensor:

$$F_{ij} = \frac{\partial x_i}{\partial x'_j}$$

Where. Fij is the deformation gradient tensor. We know by the rules of index notation that F is a second order tensor, since it has two independent indices. It is also important to note that F IS NOT Symmetric. We can also write the deformation gradient tensor in matrix format as:

$$F_{ij} = \begin{bmatrix} F_{11} & F_{12} & F_{13} \\ F_{21} & F_{22} & F_{23} \\ F_{31} & F_{32} & F_{33} \end{bmatrix} = \begin{bmatrix} \frac{\partial x_1}{\partial x'_1} & \frac{\partial x_1}{\partial x'_2} & \frac{\partial x_1}{\partial x'_3} \\ \frac{\partial x_2}{\partial x'_1} & \frac{\partial x_2}{\partial x'_2} & \frac{\partial x_2}{\partial x'_3} \\ \frac{\partial x_3}{\partial x'_1} & \frac{\partial x_3}{\partial x'_2} & \frac{\partial x_3}{\partial x'_3} \end{bmatrix}$$

Displacement Vector and Deformation Gradient Tensor

We now have two entities that relate quantities in the reference configuration to the deformed configuration: the displacement vector and the deformation gradient tensor. The next logical question is whether these quantities themselves have any connection. First, let us begin with the displacement vector that relates a position in the reference configuration to a position in the deformed configuration:

$$x_i = x'_i + u_i$$

The above equation actually represents the relationship between the three coordinates in a reference and deformed configuration:

$$x_1 = x'_1 + u_1$$

$$x_2 = x'_2 + u_2$$

$$x_3 = x'_3 + u_3$$

Now let us differentiate each on of the above equations with repect to x_1', x_2', and x_3':

$$\frac{\partial x_1}{\partial x_1'} = \frac{\partial x_1'}{\partial x_1'} + \frac{\partial u_1}{\partial x_1'}, \quad \frac{\partial x_1}{\partial x_2'} = \frac{\partial x_1'}{\partial x_2'} + \frac{\partial u_1}{\partial x_2'}, \quad \frac{\partial x_1}{\partial x_3'} = \frac{\partial x_1'}{\partial x_3'} + \frac{\partial u_1}{\partial x_3'}$$

$$\frac{\partial x_2}{\partial x_1'} = \frac{\partial x_2'}{\partial x_1'} + \frac{\partial u_2}{\partial x_1'}, \quad \frac{\partial x_2}{\partial x_2'} = \frac{\partial x_2'}{\partial x_2'} + \frac{\partial u_2}{\partial x_2'}, \quad \frac{\partial x_2}{\partial x_3'} = \frac{\partial x_2'}{\partial x_3'} + \frac{\partial u_2}{\partial x_3'}$$

$$\frac{\partial x_3}{\partial x_1'} = \frac{\partial x_3'}{\partial x_1'} + \frac{\partial u_3}{\partial x_1'}, \quad \frac{\partial x_3}{\partial x_2'} = \frac{\partial x_3'}{\partial x_2'} + \frac{\partial u_3}{\partial x_2'}, \quad \frac{\partial x_3}{\partial x_3'} = \frac{\partial x_3'}{\partial x_3'} + \frac{\partial u_3}{\partial x_3'}$$

Let's first look at and group the left hand side of the differentiated displacement relationship. If we group all these terms, we end up with the deformation gradient tensor. Now, let's take a look at the first term on the right hand side in each of the above equations. If we group these terms in a matrix, we obtain:

$$\begin{bmatrix} \frac{\partial x_1'}{\partial x_1'} & \frac{\partial x_1'}{\partial x_2'} & \frac{\partial x_1'}{\partial x_3'} \\ \frac{\partial x_2'}{\partial x_1'} & \frac{\partial x_2'}{\partial x_2'} & \frac{\partial x_2'}{\partial x_3'} \\ \frac{\partial x_3'}{\partial x_1'} & \frac{\partial x_3'}{\partial x_2'} & \frac{\partial x_3'}{\partial x_3'} \end{bmatrix}$$

Let us look more carefully at each entry in the above matrix. Obviously, if we differentiate something with respect to itself we get 1. Thus, the diagonal of the above matrix will all have ones. For all the off axis terms, we are differentiating a one coordinate position with respect to one of the two other coordinate positions. Since one coordinate position (e.g. 1,2,3) does not depend on the other coordinate positions, these terms are all zero. This leaves us with the identity matrix, which is equivalent to the kronecker delta 2nd order tensor we defined in the section on mathematical preliminaries:

$$\begin{bmatrix} 1 & 0 & 0 \\ 0 & 1 & 0 \\ 0 & 0 & 1 \end{bmatrix} = \delta_{ij}$$

Finally, we consider the term on the right hand side of the differentiated displacement equation. We can also write this in a matrix from as:

We can also recognize from the mathematical preliminary section that the above entity is the gradient of the displacement. We can therefore recognize that differentiating the displacement equation with respect to x_1', x_2', and x_3' gives us the deformation gradient tensor being equal to the kronecker delta second order tensor plus the graident gradient of the displacement. This relationship is written in both matrix and index notation below:

Matrix Notation:

$$\begin{bmatrix} F_{11} & F_{12} & F_{13} \\ F_{21} & F_{22} & F_{23} \\ F_{31} & F_{32} & F_{33} \end{bmatrix} = \begin{bmatrix} \frac{\partial x_1'}{\partial x_1'} & \frac{\partial x_1'}{\partial x_2'} & \frac{\partial x_1'}{\partial x_3'} \\ \frac{\partial x_2'}{\partial x_1'} & \frac{\partial x_2'}{\partial x_2'} & \frac{\partial x_2'}{\partial x_3'} \\ \frac{\partial x_3'}{\partial x_1'} & \frac{\partial x_3'}{\partial x_2'} & \frac{\partial x_3'}{\partial x_3'} \end{bmatrix} = \begin{bmatrix} 1 & 0 & 0 \\ 0 & 1 & 0 \\ 0 & 0 & 1 \end{bmatrix} + \begin{bmatrix} \frac{\partial u_1}{\partial x_1'} & \frac{\partial u_1}{\partial x_2'} & \frac{\partial u_1}{\partial x_3'} \\ \frac{\partial u_2}{\partial x_1'} & \frac{\partial u_2}{\partial x_2'} & \frac{\partial u_2}{\partial x_3'} \\ \frac{\partial u_3}{\partial x_1'} & \frac{\partial u_3}{\partial x_2'} & \frac{\partial u_3}{\partial x_3'} \end{bmatrix}$$

Index Notation:

$$F_{ij} = \frac{\partial x_i'}{\partial x_j'} + \frac{\partial u_i}{\partial x_j'} = \delta_{ij} + \frac{\partial u_i}{\partial x_j'}$$

Decomposing the Deformation Gradient Tensor in Stretch and Rotation Tensors

As we noted earlier, the deformation gradient tensor includes both rigid body and deformation modes. Therefore, we should be able to decompose the deformation gradient into rigid body and deformation components. Since translation does not change any vector or its components, we can therefore conclude that the deformation gradient tensor contains only deformation plus rotation components. To verify this, consider pure translation. In this case, we may write directly x in terms of x' plus a translational component:

Translation in 1, 2, and 3 denoted by s :

$$x_1 = x_1' + s_1$$
$$x_2 = x_2' + s_2$$
$$x_3 = x_3' + s_3$$

If we plug the above translation component into the deformation gradient tensor definition we have:

$$F_{ij} = \begin{bmatrix} \frac{\partial x_1'}{\partial x_1'} + \frac{\partial s_1}{\partial x_1'} & \frac{\partial x_1'}{\partial x_2'} + \frac{\partial s_1}{\partial x_2'} & \frac{\partial x_1'}{\partial x_3'} + \frac{\partial s_1}{\partial x_3'} \\ \frac{\partial x_2'}{\partial x_1'} + \frac{\partial s_2}{\partial x_1'} & \frac{\partial x_2'}{\partial x_2'} + \frac{\partial s_2}{\partial x_2'} & \frac{\partial x_2'}{\partial x_3'} + \frac{\partial s_2}{\partial x_3'} \\ \frac{\partial x_3'}{\partial x_1'} + \frac{\partial s_3}{\partial x_1'} & \frac{\partial x_3'}{\partial x_2'} + \frac{\partial s_3}{\partial x_2'} & \frac{\partial x_3'}{\partial x_3'} + \frac{\partial s_3}{\partial x_3'} \end{bmatrix}$$

In pure translation, all points in the body will move the same amount in each of the three orthogonal directions. This means that s1, s2, and s3 will not differ at different points in the body. Thus, the gradients of s1, s2 and s3 with respect to x' will all be zero. This leaves us with the fact that the deformation gradient tensor is the identity under pure translation:

$$F_{ij} = \begin{bmatrix} 1 & 0 & 0 \\ 0 & 1 & 0 \\ 0 & 0 & 1 \end{bmatrix}$$

We also not that when there is no displacement at all the deformation gradient tensor will also be equal to the identity tensor. Thus, the deformation gradient tensor does not distinguish between states of translation.

Thus, we know that the deformation gradient tensor will only contain the rigid body mode of rotation in addition to stretch. We therefore postulate without proof that the deformation gradient tensor can be decomposed into a rotation multiplied by stretch or a stretch multiplied by rotation. Let us denote the rotation tensor by R. If we decompose the deformation gradient tensor such that we rotate first and then stretch, we denote this stretch tensor by U. We therefore can write this decomposition as:

$$F_{ij} = R_{ik} U_{kj}$$

If we first stretch and then rotate, we denote the stretch tensor by V, and can write the decomposition as:

$$F_{ij} = V_{ik} R_{kj}$$

Graphically, we can represent this decomposition as:

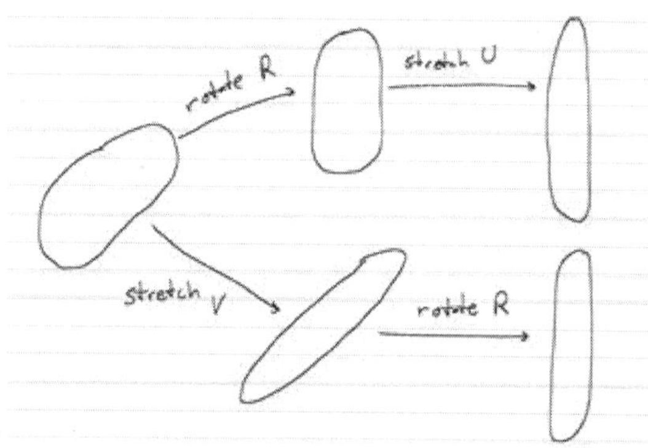

Fig Decomposition

We further note that the rotation tensor has special properties, namely that it is orthogonal. This means that the transpose of the rotation tensor multiplied by itself will give the identity tensor:

$$R^T R = I;\ R_{ki}R_{kj} = \delta_{ij}$$

This property will play a role in the definition of other deformation tensors, which themselves play a significant role in definition of many constitutive equations.

8.4.4 Large Deformation or Finite Strain Tensor

Once we have defined reference configuration, deformed configuration, displacement, and deformation gradient tensor, we can derive the large strain tensor. We cannot use the deformation gradient tensor as a strain measure directly because it can contain rigid body modes like translation and rotation. For a strain tensor, our goal is to obtain some measure the change in length of a material. One way to measure length is to take the dot product of the material vector we used when defining the deformation gradient with itself. Thus, the length squared ds'^2 of the material vector in the reference configuration is:

$$(ds')^2 = dx' \cdot dx' = dx'_i dx'_i = dx'_1 dx'_1 + dx'_2 dx'_2 + dx'_3 dx'_3$$

We can likewise write the length squared of the material vector ds^2 in the deformed configuration as:

$$(ds)^2 = dx \cdot dx = dx_i dx_i = dx_1 dx_1 + dx_2 dx_2 + dx_3 dx_3$$

The strain tensor should be a measure of how much the material vector in the reference configuration is stretched into the deformed configuration. We would essentially like the strain measure to operate on the material vector in the reference configuration and generate the difference in the square of the lengths of the material vector. We may write this idea in compact index notation as:

$$(ds)^2 - (ds')^2 = dx_i dx_i - dx'_i dx'_i = 2dx'_i E_{ij} dx'_j$$

Note that the above equation represents the difference two scalars, the square of the length of material vectors in the reference and deformed configuration. If we expand this equation we can see this more clearly:

$$\begin{aligned}
& dx_1 dx_1 + dx_2 dx_2 + dx_3 dx_3 - dx'_1 dx'_1 + dx'_2 dx'_2 + dx'_3 dx'_3 = \\
& 2dx'_1 E_{11} dx'_1 + 2dx'_1 E_{12} dx'_2 + 2dx'_1 E_{13} dx'_3 + \\
& 2dx'_2 E_{21} dx'_1 + 2dx'_2 E_{22} dx'_2 + 2dx'_2 E_{23} dx'_3 + \\
& 2dx'_3 E_{31} dx'_1 + 2dx'_3 E_{32} dx'_2 + 2dx'_3 E_{33} dx'_3
\end{aligned}$$

Note that E, which will be defined as the large deformation or finite strain tensor, has 9 components in 3D making it a 2nd order tensor. Two thirds of the strain equation above is written in terms of dx'. Therefore, to define Eij we will need to write the square of the length of the material vector in the deformed configuration also in terms of the reference configuration vector dx'. One way we can do this is to use the deformation gradient tensor, which gives the material vector dx in the deformed configuration in terms of the material vector dx' in the reference configuration.

We also need to write the relationship using the deformation gradient tensor such that we obtain pre and post multiplication by dx'i and dx'j, respectively. First, since the quantity dxidxi is a scalar with repeated or dummy indices, it doesn't matter if we write the repeated index with i,j,k, or any other letter, it will produce the same result. So, let us write the dot product with a k:

$$dx_i dx_i = dx_k dx_k$$

Next, we can write the deformed configuration material vector in terms of the deformation gradient tensor and the reference configuration material vectors as:

$$dx_k = F_{kj} dx'_j$$

This gives us the dx'j post multiplication term. Note that if we change the order of F and dx', this will not change the result. Also, since j is a repeated or dummy index in the above equation, we can replace it with any other repeated index, including i. If we carry out these operations, we obtain:

$$dx_k = dx'_i F_{ki}$$

Just to show that the Fkjdx'j and the dx'iFki give equivalent expressions, we will expand each expression. For $dx_k = dx'_i F_{ki}$

$$dx_k = F_{kj} dx'_j \Rightarrow$$

$$k = 1 \; sum \; over \; j = 1,2,3: dx_1 = \underset{j=1}{F_{11} dx'_1} + \underset{j=2}{F_{12} dx'_2} + \underset{j=3}{F_{13} dx'_3}$$

$$k = 2 \; sum \; over \; j = 1,2,3: dx_2 = \underset{j=1}{F_{21} dx'_1} + \underset{j=2}{F_{22} dx'_2} + \underset{j=3}{F_{23} dx'_3}$$

$$k = 3 \; sum \; over \; j = 1,2,3: dx_3 = \underset{j=1}{F_{31} dx'_1} + \underset{j=2}{F_{32} dx'_2} + \underset{j=3}{F_{33} dx'_3}$$

For $dx_k = dx'_i F_{ki}$ we have:

$$dx_k = dx'_i F_{ki} \Rightarrow$$
$$k=1\ sum\ over\ i=1,2,3: dx_1 = \underset{i=1}{dx'_1 F_{11}} + \underset{i=2}{dx'_2 F_{12}} + \underset{i=3}{dx'_3 F_{13}}$$
$$k=2\ sum\ over\ i=1,2,3: dx_2 = \underset{i=1}{dx'_1 F_{21}} + \underset{i=2}{dx'_2 F_{22}} + \underset{i=3}{dx'_3 F_{23}}$$
$$k=3\ sum\ over\ i=1,2,3: dx_3 = \underset{i=1}{dx'_1 F_{31}} + \underset{i=2}{dx'_2 F_{32}} + \underset{i=3}{dx'_3 F_{33}}$$

we can see that the two above equations are equivalent. We can now substitute the expressions with the deformation gradient into the strain equation to obtain:

$$dx'_i F_{ki} F_{kj} dx'_j - dx'_i dx'_i = 2 dx'_i E_{ij} dx'_j$$

The last task in defining Eij is to somehow rewrite the dx'idx'i term as a dx'idx'j term. To do this, we can simply apply the Kronecker delta as:

$$dx'_i = \delta_{ij} dx'_j$$

We can expand the above relationship as:

$$i=1: dx'_1 = \delta_{11} dx'_1 + \delta_{12} dx'_2 + \delta_{13} dx'_3 = 1dx'_1 + 0dx'_2 + 0dx'_3 = dx'_1$$
$$i=2: dx'_2 = \delta_{21} dx'_1 + \delta_{22} dx'_2 + \delta_{23} dx'_3 = 0dx'_1 + 1dx'_2 + 0dx'_3 = dx'_2$$
$$i=3: dx'_3 = \delta_{31} dx'_1 + \delta_{32} dx'_2 + \delta_{33} dx'_3 = 0dx'_1 + 0dx'_2 + 1dx'_3 = dx'_3$$

which shows that multiplying by the kronecker delta gives us back the original quantity. We can now substitute in for the kronecker delta term into the large strain equation to give:

$$dx'_i F_{ki} F_{kj} dx'_j - dx'_i \delta_{ij} dx'_j = 2 dx'_i E_{ij} dx'_j$$

We note that the above relationship can be separated in terms of dx'i as:

$$dx'_i \left(F_{ki} F_{kj} - \delta_{ij} - 2E_{ij} \right) dx'_j = 0$$

since both dx'i and dx'j are non-zero and arbitrary, then the quantity in the bracket must be zero. This gives:

$$F_{ki} F_{kj} - \delta_{ij} - 2E_{ij} = 0 \Rightarrow$$
$$2E_{ij} = F_{ki} F_{kj} - \delta_{ij}$$

We divide by two to come up with our final definition of the finite strain tensor in terms of the deformation gradient tensor as:

$$E_{ij} = \frac{1}{2}\left(F_{ki}F_{kj} - \delta_{ij}\right)$$

We will expand the above expression for one normal and one shear stress. First, consider the normal stress E11. In this case, i and j are fixed at 1. However, k in the above is a repeated or dummy index, and therefore terms must be summed over k. Thus, we can write:

$$i=1; j=1:$$

$$E_{11} = \frac{1}{2}(F_{k1}F_{k1} - \delta_{11}) = \frac{1}{2}(F_{k1}F_{k1} - 1)$$

$$sum\ over\ k = 1,2,3$$

$$E_{11} = \frac{1}{2}(F_{11}F_{11} + F_{21}F_{21} + F_{31}F_{31} - 1)$$

We also show an example of the shear strain E12:

$$i=1; j=2:$$

$$E_{12} = \frac{1}{2}(F_{k1}F_{k2} - \delta_{12}) = \frac{1}{2}(F_{k1}F_{k2} - 0)$$

$$sum\ over\ k = 1,2,3$$

$$E_{12} = \frac{1}{2}(F_{11}F_{12} + F_{21}F_{22} + F_{31}F_{32})$$

In addition to the finite strain tensor, other deformation tensors are often defined in terms of the deformation gradient tensor. An often used deformation measure, especially in hyper elastic constitutive tensors used to characterize soft tissues, is the right Cauchy deformation tensor. This is defined as:

$$C_{ij} = F_{ki}F_{kj}$$

The right Cauchy deformation tensor can also be defined in matrix form as:

$$C = F^T F$$

We can expand both the index and matrix notation of the right Cauchy deformation tensor as:

$$C_{ij} = F_{ki}F_{kj}$$
$$i=1, j=1: C_{11} = F_{11}F_{11} + F_{21}F_{21} + F_{31}F_{31}$$
$$i=1, j=2: C_{12} = F_{11}F_{12} + F_{21}F_{22} + F_{31}F_{32}$$
$$i=1, j=3: C_{11} = F_{11}F_{13} + F_{21}F_{23} + F_{31}F_{33}$$
$$i=2, j=1: C_{21} = F_{12}F_{11} + F_{22}F_{21} + F_{32}F_{31}$$
$$i=2, j=2: C_{22} = F_{12}F_{12} + F_{22}F_{22} + F_{32}F_{32}$$
$$i=2, j=3: C_{23} = F_{12}F_{13} + F_{22}F_{23} + F_{32}F_{33}$$
$$i=3, j=1: C_{31} = F_{13}F_{11} + F_{23}F_{21} + F_{33}F_{31}$$
$$i=3, j=2: C_{32} = F_{13}F_{12} + F_{23}F_{22} + F_{33}F_{32}$$
$$i=3, j=3: C_{33} = F_{13}F_{13} + F_{23}F_{23} + F_{33}F_{33}$$
$$C = F^T F$$
$$F = \begin{bmatrix} F_{11} & F_{12} & F_{13} \\ F_{21} & F_{22} & F_{23} \\ F_{31} & F_{32} & F_{33} \end{bmatrix}; F^T = \begin{bmatrix} F_{11} & F_{21} & F_{31} \\ F_{12} & F_{22} & F_{32} \\ F_{13} & F_{23} & F_{33} \end{bmatrix}$$
$$C = \begin{bmatrix} F_{11} & F_{21} & F_{31} \\ F_{12} & F_{22} & F_{32} \\ F_{13} & F_{23} & F_{33} \end{bmatrix} \begin{bmatrix} F_{11} & F_{12} & F_{13} \\ F_{21} & F_{22} & F_{23} \\ F_{31} & F_{32} & F_{33} \end{bmatrix} =$$
$$\begin{bmatrix} F_{11}F_{11} + F_{21}F_{21} + F_{31}F_{31} & F_{11}F_{12} + F_{21}F_{22} + F_{31}F_{32} & F_{11}F_{13} + F_{21}F_{23} + F_{31}F_{33} \\ F_{12}F_{11} + F_{22}F_{21} + F_{32}F_{31} & F_{12}F_{12} + F_{22}F_{22} + F_{32}F_{32} & F_{12}F_{13} + F_{22}F_{23} + F_{32}F_{33} \\ F_{13}F_{11} + F_{23}F_{21} + F_{33}F_{31} & F_{13}F_{12} + F_{23}F_{22} + F_{33}F_{32} & F_{13}F_{13} + F_{23}F_{23} + F_{33}F_{33} \end{bmatrix}$$

By closely examining the explicit expression for C, we can see that C is symmetric, that is Cij = Cji. In addition to the right Cauchy deformation tensor, we can define the left Cauchy deformation tensor in both index and matrix form as:

$$B_{ij} = F_{ik}F_{jk}$$
$$B = FF^T$$

We can similarly expand the left Cauchy deformation tensor as:

$$B_{ij} = F_{ik}F_{jk}$$
$$i=1, j=1: B_{11} = F_{11}F_{11} + F_{12}F_{12} + F_{13}F_{13}$$
$$i=1, j=2: B_{12} = F_{11}F_{21} + F_{12}F_{22} + F_{13}F_{23}$$
$$i=1, j=3: B_{11} = F_{11}F_{31} + F_{12}F_{32} + F_{13}F_{33}$$
$$i=2, j=1: B_{21} = F_{21}F_{11} + F_{22}F_{12} + F_{23}F_{13}$$
$$i=2, j=2: B_{22} = F_{21}F_{21} + F_{22}F_{22} + F_{23}F_{23}$$
$$i=2, j=3: B_{23} = F_{21}F_{31} + F_{22}F_{32} + F_{23}F_{33}$$
$$i=3, j=1: B_{31} = F_{31}F_{11} + F_{32}F_{12} + F_{33}F_{13}$$
$$i=3, j=2: B_{32} = F_{31}F_{21} + F_{32}F_{22} + F_{33}F_{23}$$
$$i=3, j=3: B_{33} = F_{31}F_{31} + F_{32}F_{32} + F_{33}F_{33}$$

From the right and left Cauchy deformation tensors, we can further define relationships between these tensors and the stretch tensors. First, consider the definition of the right Cauchy:

$$C_{ij} = F_{ki} F_{kj}$$

We next substitute the polar decomposition of F for each F in the above equation:

$$F_{ki} = R_{kl} U_{li} ; F_{kj} = R_{km} U_{mj}$$

$$C_{ij} = F_{ki} F_{kj} = R_{kl} U_{li} R_{km} U_{mj}$$

However, we know that since R is an orthogonal tensor, we have the following relationship:

$$R_{kl} R_{km} = \delta_{lm} \Rightarrow$$

$$C_{ij} = U_{li} \delta_{lm} U_{mj} = U_{li} U_{lj}$$

This result shows that the right Cauchy deformation tensor is equal to the square of the right stretch tensor. Likewise, we can show that the left Cauchy Deformation tensor is:

$$B_{ij} = F_{ik} F_{jk} \Rightarrow$$

$$F_{ik} = V_{il} R_{lk} ; F_{jk} = V_{jm} R_{mk}$$

$$B_{ij} = V_{il} R_{lk} R_{mk} V_{jm}$$

$$B_{ij} = V_{il} \delta_{lm} U_{jm} = V_{il} V_{jl}$$

Thus giving that the left Cauchy Deformation tensor is the square of the left stretch tensor.

We can furthermore find the eigen-values of both U and V. The eigen-values of U are called the principal stretches. These are used to construct some constitutive equations.

8.5 Cardiovascular Mechanics

8.5.1 Introduction

The cardiovascular system has three main components:

a. Blood is the medium of convective transport of oxygen, carbon dioxide, nutrients, and other solutes between the organs, followed by diffusive transport to the tissues at cellular level. Without this convective transport an appropriate exchange of the solutes would be impossible due to a too large diffusional resistance

b. The circulatory system is the distribution system for blood and is made up of a network of blood vessels, namely the arteries which transport blood away from the heart and the veins, which the veins which transport towards the heart.

c. The heart is a four-chambered pump that drives the blood flow in the circulatory system. The circulatory system may be subdivided into two systems in series, the pulmonary circulation and the systemic circulation as shown in figure.

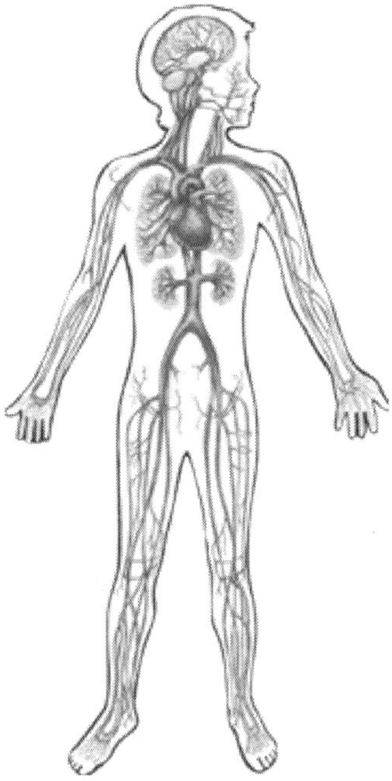

Fig pulmonary circulation and the systemic circulation

Blood flows from the superior and inferior vena cava in to the right atrium, then through the tricuspid valve in to the right ventricle, subsequently through the semi-lunar valves into the pulmonary artery, which strongly bifurcates in pulmonary arterioles transporting the blood to the lungs. Oxygenated blood is the transported through the pulmonary veins to the left atrium, and further through the mitral valve, the left ventricle, and finally thorough the aortic valve into the aorta. The peripheral circulation starts with the aorta, perfuses various organs, and returns to the right atrium. In each organ, flow begins in the arteries, perfuses the micro-circulatory bed, and finally drains into veins. The vena cava drain blood from the various organs, and return it to the heart.

The cardio-vascular (CV) system generally consists of heart (cardio) and blood vessels (vascular). Specifically, the CV system can be viewed as the pulmonary circulation (composed of the right heart and the lungs) and the systemic circulation in which the left heart supplies blood to the systemic organs (heart muscle, brain, skeletal muscle, bone, gastrointestinal system, spleen, liver, kidney, skin and others). The pulmonary and systemic circulations are arranged in series and hence the right and left heart chambers must pump identical volumetric flow rate in health.

The primary function of the CV system is to circulate the blood through the capillaries to within diffusion distances (less than or equal to 10 μm) of every tissue parenchyma. The blood serves as the vehicle for delivery and removal of nutrients and waste products, respectively, and the vessels serve as the highways. The blood vessels are not merely highways or pipe channels, however, as they have a passive capacitance function that maintains the blood pressure in diastole as well as an active auto-regulatory control that allows the organ to respond to local demands.

Arterial anatomy

Inspired by the presentation, an overview of the of the arterial anatomy is given in this section. The aorta is the main artery with a proximal diameter of 2-3 cm in the human adult. Originating from the left ventricle, the aorta is separated from the ventricle lumen during diastole, whereas there is direct communication between the left ventricle and the aorta during systole.

The aorta is the most proximal artery connected directly to the heart and acts both as a conduit and an elastic chamber. In its latter role, the aorta's elasticity serves to convert the heart's pulsatile flow to nearly steady flow in peripheral vessels. Stephen Hales (1733) reported his observations that the aorta expands to accommodate a large fraction of the stroke volume. Hales reasoned that the aorta serves as an elastic reservoir, which distends with blood during the heart's contraction and discharges blood through the peripheral resistance by elastic recoil while the heart refills. His explanation accounts for the smoothing action of the aorta in converting the pulsatile flow of the heart to smooth flow in blood vessels. An analogy was made with a hand-pump fire engine which smoothes a pump's pulsatile flow of water.

The aorta tapers and the distal aortic diameter is about 1 cm. Normally, three main branches are found on the top of the aortic arch, bifurcating into the arteries towards the head and the upper limbs. The descending aorta is divided into the thoracic and the abdominal part as the aorta crosses the diaphragm. In the abdomen, important aortic branches provide blood flow to the kidneys (constituting about 15\% of the blood volume) and the gastro-intestinal organs. At the distal end, the aorta bifurcates into two iliac arteries, providing blood supply to the lower extremities.

Towards the head, the most important branches are the left and right common carotid arteries, which bifurcate further into the internal and external carotid arteries. These arteries are superficial, and are therefore readily accessible for noninvasive measurements of pressure and flow. The pressure pulsation can easily be felt in then neck. Blow supply to the upper limbs are provided by the subclavian artery, which changes name to the brachial artery in the upper arm, and bifurcates at the elbow to the radial and the ulnar artery. These arteries are the most commonly used sites for non-invasive pressure recordings: the brachial for sphygmomanometric measurements; the radial and subclavian arteries for tonometric measurements.

In the abdomen, the most important branches are those which supply the gastro-intestinal organs with blood; hepatic artery to the liver; gastric artery to the stomach, splenic artery to the spleen, the superior and inferior mesenteric arteries to the intestines, and the renal arteries to the kidneys. The renal arteries captures up to 15\% of the total stroke volume, whereas all abdominal arteries receives about 40\% of the total stroke volume.

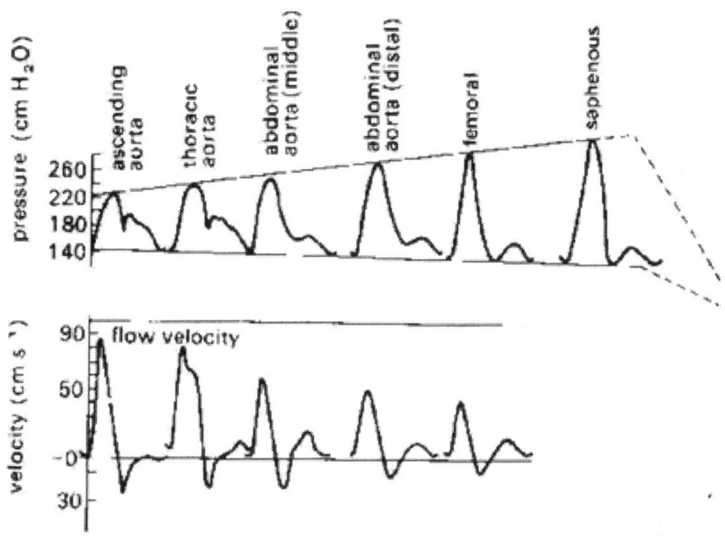

During systole, the ventricle ejects blood into the arteries, giving rise to pressure and flow waves traveling from the heart towards the periphery. The pressure peak is delayed with increasing distance from the heart, indicating wave propagation along the aorta with a finite pulse wave velocity or wave speed. Further, the pressure wave is characterized by a progressively increasing amplitude (systolic pressure), and a steepening of the front. Only a moderate fall in the mean pressure with increasing distance is observed, indicating minor effects of viscous friction. The viscous effects are limited to only a few mmHg in the large and mid-sized arteries (e.g., radial artery).

The rise in the pressure amplitude on the other hand, is determined by the varying local characteristic impedance (i.e., elastic properties of the arteries and the vessel dimensions), as well as heart rate and stroke volume. As we will learn in more detail in the chapter Blood flow in compliant vessels; any change in the characteristic impedance, which depends on the local pulse wave velocity (or compliance) and the cross-sectional area of the vessels, will give rise to wave reflections.

For now we settle for the explanation that the observed wave phenomenon in the arterial tree is a direct consequence of the locally varying distensibility or compliance of the arterial wall, allowing for a partial and temporary storage of the blood ejected from the heart. In accordance with the increase in pressure amplitude, the flow wave is damped progressively due to the reflections, and to the buffering capacity (or compliance) of the arteries towards the periphery. In total, the pulsatile flow in the aorta is transformed to a quasi stationary flow at the arteriolar and capillary level.

Chapter 9
Applying Biomechanical Principles

9.1 Biomechanical Analysis

Biomechanics is the study of human motion. The study of biomechanics is important when determining what causes injuries and therefore how we can prevent them re-occurring.

Biomechanical analysis can benefit athletes of all ages and skill levels. Whether you are a weekend runner or a high performance sprinter, biomechanical analysis can be beneficial to you as it will allow you the opportunity to develop more efficient movement patterns. This can benefit the weekend runner by allowing them to increase their distance and run pain free, while the high performance sprinter can benefit from a more efficient running stride and allow them to shave milliseconds of off their personal best time. This is especially important in elite athletes but can be a major cause in recurrent injuries in the less gifted amateur athlete. Biomechanical analysis can involve:

- gait analysis - study of your walking pattern.
- running analysis - study of your running style.
- video analysis or motion capture analysis.
- sports biomechanics - sport specific analysis.
- workplace analysis - study of how you do you job.
- biomechanics of running, sprinting, swimming, throwing etc.

For the purpose of understanding the movement being analyzed, strength and conditioning professionals should use the following four questions. First, what are the patterns of movement (i.e., concentric, eccentric, or isometric), and in which planes do they take place? Second, what joints are involved during the activity? Third, what muscles are recruited, and what are the muscle actions?

Finally, what is the duration of time that the athlete will be actively engaged in the athletic event? With these key questions, strength and conditioning professionals can determine the demands placed on the body during the sport. The ultimate goal of analysis is to manipulate and match the acute variables that govern the program's design to match the metabolism and movements involved in the sport.

Typically, biomechanical evaluations require strength and conditioning professionals to analyze videos of athletes performing their sports. Those without access to advanced video equipment can accomplish this type of analysis by watching simple video of athletes during practices or games.

9.2 Biomechanics and Training

Biomechanics is the science of movement of a living body, including how muscles, bones, tendons and ligaments work together to produce movement. Biomechanics is part of the larger field of kinesiology, specifically focusing on the mechanics of movement. It is both a basic and applied science, encompassing research and practical use of its findings.

Biomechanics includes not only the structure of bones and muscles and the movement they can produce, but also the mechanics of blood circulation, renal function, and other body functions. The American Society of Biomechanics says that biomechanics represents the broad interplay between mechanics and biological systems. Biomechanics studies not only the human body but also animals and even extends to plants and the mechanical workings inside cells.

Elements of Biomechanics

- Statics: Studying systems that are in equilibrium, either at rest or moving at a constant velocity.
- Dynamics: Studying systems that are in motion with acceleration and deceleration.
- Kinematics: Describing the effect of forces on a system, motion patterns including linear and angular changes in velocity over time. Position, displacement, velocity and acceleration are studied.
- Kinetics: Studying what causes motion, the forces and moments at work.
- Sports Biomechanics
- Sports biomechanics studies human motion during exercise and in sports. Physics and the laws of mechanics are applied to athletic performance.

Biomechanics can be applied to individuals, analyzing their movements and coaching them for more effective movement during exercise and sports movement. An individual's running gait or golf swing can be filmed and recommendations made for them to change and improve it.

- Biomechanics can be used in the design of sports equipment, clothing, shoes, and the fields and facilities where sports are played. A shoe can be designed for the best performance for a middle distance runner or a racket for the best grip.
- Biomechanics can study sports techniques and training systems and develop ways to make them more efficient. This can include basic research into how hand position affects propulsion in swimming. It can propose and analyze

new training techniques based on the mechanical demands of the sport, aimed at resulting better performance.

❖ Biomechanics can be applied to studying the causes, treatment, and prevention of sports injuries. Research analyzes the forces at work that can lead to an ankle sprain and how shoe design or the playing surface might reduce the risk of injury.

9.3 Mechanical Stress and Injury

Biomechanics is concerned with the internal and external forces that act on the human body and the effects produced by these forces. Most biomechanics textbooks state two primary goals for biomechanics in sport and exercise training: performance enhancement and injury prevention/rehabilitation. Biomechanics enhances performance by using mechanical principles to improve an individual's technique, the equipment they use (machine engineering), and to modify specific technique and training protocols the coach or trainer implements to help an individual achieve their goals. Similarly, for injury prevention and rehabilitation, biomechanics is used to develop techniques that reduce the chance of injury.

9.3.1 Tissue Stress

The tissue stress approach to mechanical foot therapy is based on the concept that the location, type and magnitude of stresses that cause mechanical symptoms of the foot and/or lower extremity should be strongly considered when specific treatments are being designed for a patient. The ideas inherent in the tissue stress approach should not be foreign to any clinician who routinely treats mechanically based symptoms of the foot and/or lower extremity. For example, if a patient has pain plantar to the second metatarsophalangeal joint (MPJ), the tissue stress approach simply states that treatment should be directed towards reducing any abnormal stresses in the injured structures of the plantar second MPJ so that healing may be accelerated and a normal gait pattern may be reestablished. This type of approach to mechanical foot therapy has a different emphasis than that advocated by the proponents of the subtalar joint neutral (SJN) theory in which mechanical treatment is directed at preventing "abnormal compensations" for "deformities" in the foot and/or lower extremity. When prescriptions for custom foot orthoses are considered, the SJN theory places much less emphasis on the anatomical site and type of pathological stresses which are causing the symptoms and/or injury in the patient than does the tissue stress approach to mechanical foot therapy.

In order for the clinician to become more familiar with the idea of tissue stress, it is important that the basic mechanical concepts of loading forces and stresses first be reviewed. The external forces that cause stresses to occur within a structure may be classified by the way they tend to deform the structure upon which they are acting. Axial loading can occur in either one of two ways: compression and tension.

A structure is defined as being under compression if the axial loading force tends to make it shrink as shown in figure. If the axial loading force tends to make a structure elongate, then it is defined as being under tension. Another form of loading on a structure is called shear loading. Shear loading tends to produce horizontal sliding of one layer of a structure over another. Tensile and compressive forces are commonly called normal forces while shearing forces are commonly called tangential forces.

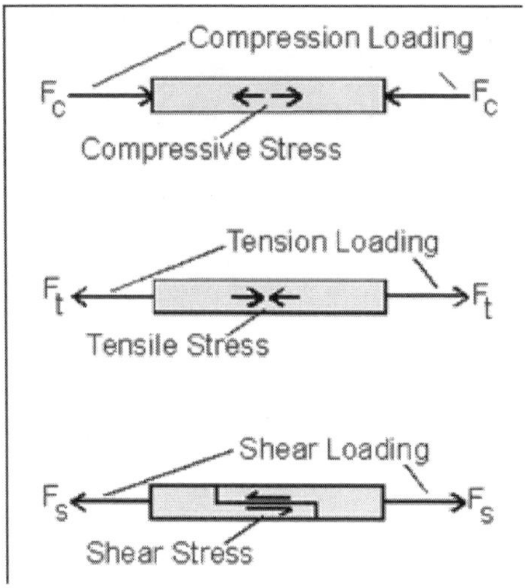

Fig External loading forces acting on structure resulting in development of internal resistance to load

Any tissue in the body, when subjected to an external loading force, will develop an internal resistance to that load, which is called stress. If a tissue tends to have a large capacity to resist external loading forces, then the internal load, or stress, it can develop under load will be relatively large. However, if an anatomical structure has only a small capacity to resist the loading forces, then the internal load, or stress, it can develop under load will be relatively small. For example, a steel cable is able to develop higher stresses within its structure when compared to a rubber band since the steel cable has a larger capacity to resist tensile loading forces than does the rubber band.

Depending on the nature of the external loading force acting on it, a structure may develop compressive stress, tensile stress and/or shear stress as shown above. In order to develop a compressive stress, the structure must develop an internal loading force that tends to resist being pushed together (e.g. the tibia develops compressive stress in response to supporting the mass of the body during weight bearing activities). In order to develop a tensile stress, the structure must develop an internal loading force that tends to resist being pulled apart (e.g. the Achilles tendon develops tensile stress in response to ankle joint dorsiflexion loads). Any forces acting parallel, or tangential, to the applied external loading force creates a shear stress. Shear stress is commonly developed in structures where torsional forces are applied to that structure. For example, the tibia will develop shear stresses directed perpendicular to its long axis when torsional forces are applied that tend to cause internal rotation of the tibia relative to the foot.

The amount of stress within a structure may be calculated by dividing the magnitude of the axial load, F, by the cross-sectional area over which the load is distributed, A. Axial, or normal, stress is commonly denoted by the Greek letter, sigma. Therefore, the formula for determining axial stress is as follows: $\sigma = F/A$. Forces that act tangential to the applied load create shear stress which is commonly denoted by the Greek letter, tau. Therefore, the formula for determining shear stress is as follows: $\tau = F/A$. The standard unit used for the measurement of stress in biomechanics is the Pascal (Pa), which is defined as one Newton (N) distributed over one square meter (1 Pa = 1 N/m2). One Newton is equivalent to 0.225 pounds.

Since the magnitude of stress acting on a biological tissue is determined by the magnitude of loading force divided by the cross-sectional area over which the load is distributed, then cross-sectional size of the tissue is just as important as the absolute force acting on the tissue. For example, if a tensile loading force of 300 N is applied to a posterior tibial tendon with a cross-sectional area of 0.25 cm2, then the stress on the posterior tibial tendon would be as follows: σ = 300 N/0.0025 m2 = 120,000 Pa.

However, if the same loading force of 300 N is applied to a posterior tibial tendon where half of the tendon fibers have been ruptured, so that the cross-sectional area of the tendon is now 0.125 cm2, then the magnitude of stress acting on the remaining tendon fibers would be increased by two-fold as follows: σ = 300 N/0.00125 m2 = 240,000 Pa. Therefore, the loss or rupture of a percentage of fibers in a tendon can greatly increase the stress on the remaining tendon fibers that, in turn, will increase the likelihood of tendon rupture even under normal magnitudes of tensile loading forces on the tendon.

9.4 Tissue Response to Stress

9.4.1 Introduction

In terms of weight and volume, water is indisputably the major component of soft biological tissues. The mechanical property most commonly associated with the presence of water is near volume constancy, i.e., incompressibility, based on the fact that the compression modulus of water is orders of magnitude larger than the distortional stiffness of tissues. The vast majority of biomechanical studies on soft collagenous tissues is based on this assumption, typically without experimental verification of its validity. While this assumption implies that all the interstitial fluid is immobile and bound to the tissue, other biomechanical analyses build, vice-versa, on the mobility of the liquid phase in a porous matrix, driven by spatial pressure gradients. The role of water in such biphasic or porous media representations thus lies in that it furnishes the tissue with time-dependent characteristics governed by the chemical and physical properties of fluid and solid phase3. The hydrostatic pressure in a material that is under uniaxial (UA) tensile stress σ is −σ/3, hence negative, and Biot's theory of poroelasticity4 tells us that such negative pressure should lead to an increase of volume under tensile loading. For a saturated porous medium, this indicates influx of water. In contrast to both conceptions, negative compressibility5 documented by Poisson's ratios larger than 0.5 and volume loss during UA tensile loading, has been reported for tendons and ligaments, human amnion, samples prepared from articular cartilage, as well as collagen and fibrin-based hydrogels. This entails efflux of water for negative hydrostatic pressure, and we refer to this behavior as inverse poroelasticity. The corresponding volume changes reported for tendons and ligaments 6,7,8 are comparably small, and substantial volume reductions in cartilage were elicited only by large elongations.

During the demonstration, various soft collagenous membranes experiences very large volume reduction for moderate in-plane tensile loading, which the volume changes are reversible, and that are collective result of particular arrangement and kinematics of collagen network and the inherent chemo elasticity of soft biological tissues. While the important role of water in creating residual stresses in tissues by swelling has been identified, and its impact on the mechanics of collagen on a molecular scale was recently highlighted, the present data demonstrate the key role of water mobility in determining the tissue response, and identify an essential deformation mechanism with far-reaching implications for the biomechanics and mechano-biology of soft biological tissues. We show this on the basis of UA and biaxial tensile experiments on human amniotic membrane (hAM), bovine liver capsule (bGC) and porcine pericardium (pPC), i.e., three different types of biological tissues. Monotonic and cyclic loading protocols are applied in macroscopic tests and in situ, in a multi-photon microscope (MPM), allowing for a quantitative analysis of the 3D kinematics of the tissue associated with each loading condition. Drastic volume changes are associated with strong variations in water content. As a consequence, the osmotic pressure varies with mechanical loads and, vice-versa, the level of tension for a given deformation depends on the chemical potential of the interstitial fluid. The observed chemo-mechanical coupling is rationalized through corresponding biphasic constitutive models and is proposed as an effective mechanism for mechano-transduction. Inverse poro-elasticity and the associated compaction and dilation of the collagen fiber network provide soft tissues with unique mechanical properties, such as high compliance for moderate stretching and strong resistance to larger non-physiological deformations. MPM observations of the near field of crack-like defects demonstrate that these mechanisms induce a localized reversible densification of collagen protecting the tissue from crack propagation.

9.4.2 Description

The liquid volume fraction of the considered membranes is typically >80%, the network of collagen fibers (mainly types I and III) constitutes a significant amount of tissue dry weight and no significant elastin content was reported for the bulk of these membranes. Considered as a mechanical system, these tissues thus consist mainly of water within a network of collagen fibers, enriched with negatively charged proteoglycans as shown in figure.

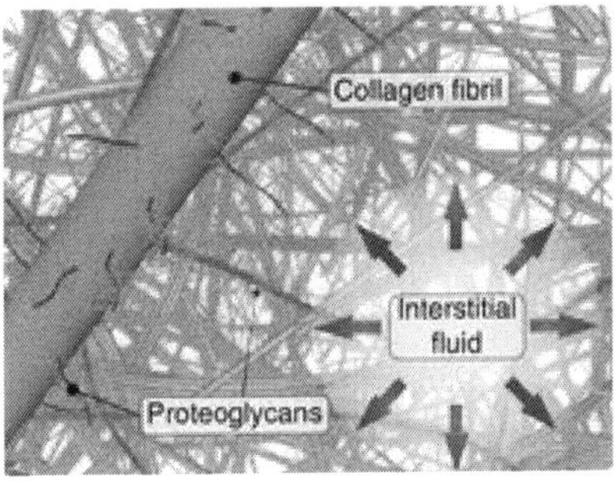

Fig Tissues composition

The present analysis of their deformation behavior is based on the kinematic response of membrane samples to UA tension and in-plane biaxial tension, performed in wet conditions. Figure below illustrates qualitatively the observed kinematic response to UA and biaxial loading common to all three membranes: large contraction perpendicular to the loading direction leads to fluid efflux and strong volume reduction upon application of tensile forces.

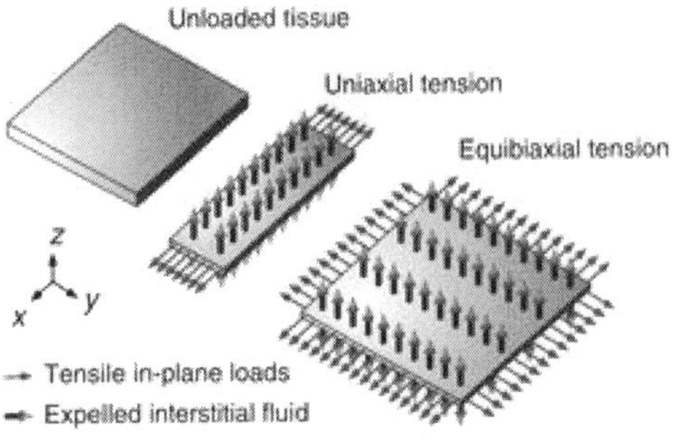

Fig qualitatively kinematic response to UA and biaxial loading of membranes

MPM images of the lateral cross sections taken at increasing values of elongation in in situ UA tensile tests highlight the drastic reduction in thickness as shown.

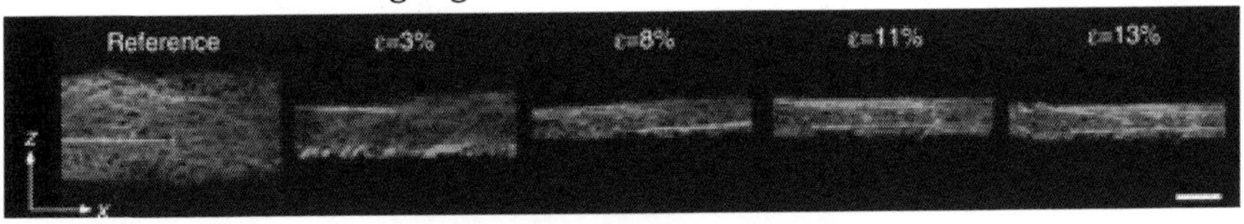

Fig Tissues tensile tests

To quantify volume changes in all three membranes, the in-plane state of deformation was retrieved from the displacement field of a surface pattern in macroscopic experiments, while MPM image stacks from in situ experiments were used to determine the corresponding changes of thickness for UA and biaxial loading conditions as shown in figure below.

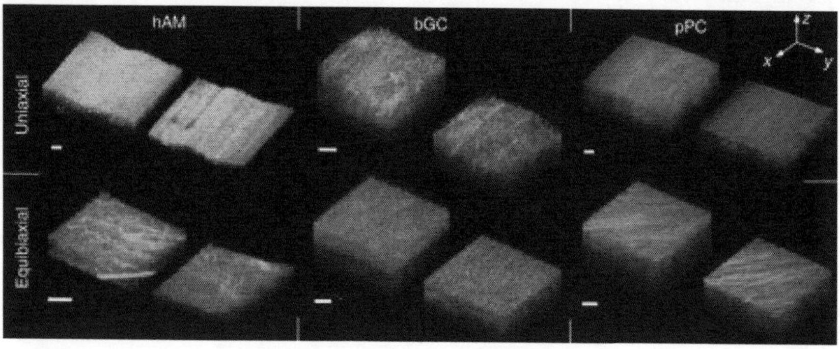

On the basis of these data, the apparent Poisson ratios and the change in volume for monotonic tensile tests were quantified as shown in figure next. Poisson's ratios change with elongation and reach maxima up to 6, about twice the values reported for biopolymer gels.

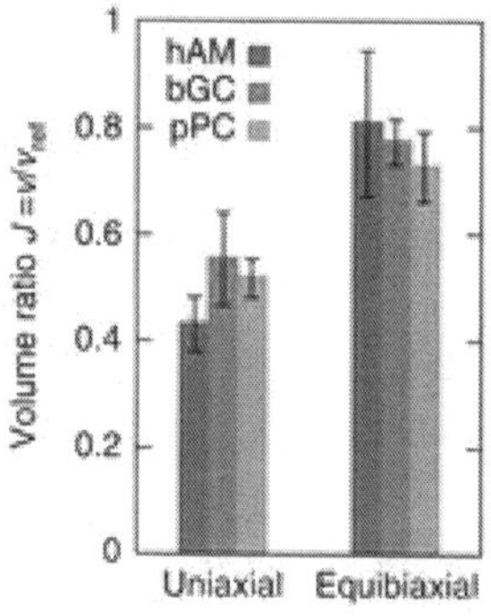

Fig monotonic tensile test

For moderate UA elongation, the volume decreases by up to a factor of two for all three membranes, and even four-fold when larger forces were applied. Less, but

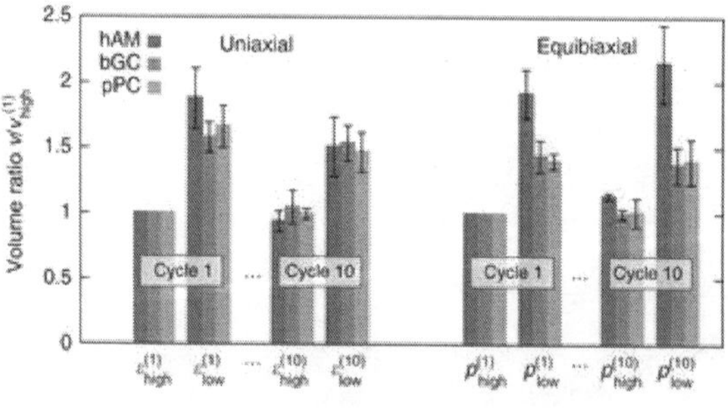

Fig

still remarkable volume reduction was associated also with equibiaxial (EB) in-plane tension.

The question arises whether the drastic volume changes are a characteristic of the very first loading step (after tissue extraction from the organism) that would accordingly not be present or, at least, much smaller in successive loading cycles. To evaluate this, tissue specimens were cyclically loaded and unloaded 10 times between two load levels, the upper one causing ~10% elongation, and the corresponding volume changes were determined. The results convincingly showed that volume loss and regain persist over the cycles, with volume changes of 50% or larger.

Volume increase and decrease hence represent a characteristic feature of the deformation behavior of all three membranes under both UA and biaxial loading conditions. In terms of resistance to deformation, all tissues display the characteristic J-shaped tension-stretch curves, well known for soft biological materials. The in-plane lateral contraction in UA tensile tests is by a factor of 5 to 10 larger than the one expected for isotropic incompressible materials and both this contraction and the volume reduction data are highly reproducible as opposed to the typically large variability of the corresponding tension-elongation curves.

9.5 Mechanism of Overuse Injury

9.5.1 Introduction

In a basic biomechanical model that takes tissue properties as well as load characteristics into account, injury results from a transfer of energy to the tissue. The mechanical properties of human tissue, such as stiffness (stress–strain relation) and ultimate strength, governs how the body responds to physical loads. They differ for each tissue and are dependent on the nature and type of load, its rate, the frequency of load repetition, the magnitude of energy transfer, and intrinsic factors such as age, sex, and physical condition. In this model, it is the relation between load and load tolerance that determines the injury outcome of an event. The key point to consider with regard to biomechanical factors is that they must explain how the event either resulted in a mechanical load in excess of that tolerated under normal circumstances or reduced the tolerance levels to a point at which a normal mechanical load cannot be tolerated.

When comparing the biomechanical model with Meeuwisse's epidemiological model, load and load tolerance can be influenced in different ways by the main elements of the epidemiological model: internal risk factors, external risk factors, and the incident event. Although all of the three elements influence load, load tolerance is mainly determined by the internal risk factors. For instance, important determinants for load tolerance, such as the material properties and size of a ligament, are influenced by age, sex, body size, and training background. The same factors can also influence load—for example, a large male footballer generates higher ligament loads than a small female gymnast performing the same task. External risk factors also influence load; protective equipment such as a helmets attenuate loads, whereas training on a hard surface increases load. In the same way, the inciting event clearly determines the load.

McIntosh21 has recently described a more complex biomechanically focused model of injury causation to account for additional factors that may influence the interplay between load and load tolerance (positively or negatively), such as behaviour/attitudes, training, skills, equipment, coaching, other competitors, and the environment. One of the purposes of this model is to describe how load and load tolerance and hence injury risk can change as a result of changes in such factors through interventions. For example, a helmet will attenuate impact energy, thereby reducing the head impact force, and skills training may enable someone to maintain their balance over the weight bearing knee, thus reducing knee loads in the frontal and transverse planes. Improved fitness may protect the tissue against injury through the effects of training on its material properties, but can also result in higher forces being applied to the tissue. If a tennis player through strength, fitness, or skill training develops a faster serve, this can result in a higher load on his/her shoulder, as well as for the opponent.

These models can also be used to account for risk homoeostasis (risk compensation), changes in behaviour resulting from the introduction of a safety measure. For example, a skier wearing a helmet may take greater risks, such as skiing faster or more aggressively, on more difficult runs, through trees, or off piste. Thus the end result of introducing a preventive measure may be no change, or even an increase in injury risk, and the need to account for such unwanted effects when introducing protective equipment has recently been emphasised.

9.5.2 Injury Mechanism

Regardless of whether a biomechanical or epidemiological model is used to describe the interaction between the different causative factors, a precise description of the inciting event is critical. The ultimate goal is to use this information to develop specific preventive measures for a specific injury type, perhaps even in a specific sport. To this end, different ways of classifying injury mechanisms have been proposed in the literature. The Committee on Trauma Research classifies causal mechanisms into:

a. crushing deformation;

b. impulsive impact;

c. skeletal acceleration;

d. energy absorption;

e. extent and rate of tissue deformation.

Another source proposes seven basic mechanisms of injury:

a. contact of impact;

b. dynamic overload;

c. overuse;

d. structural vulnerability;

e. inflexibility;

f. muscle imbalance;

g. rapid growth.

However, although these definitions may make sense from a mechanical perspective and be helpful in other areas of trauma research, we would argue that neither of these systems is particularly helpful if the objective is to identify modifiable causes with a potential for intervention in sports injuries.

If we were to describe a situation where a basketball player sustained an ACL injury, one report might be that the injury was caused by "an attacker planting and cutting to set up for a shot". This description includes aspects of the playing situation and skill performed when injured. In addition, the explanation could add that "the incident occurred to a powerful attacker, who was pushed just as she was trying to pass the opponent in a maximal effort", which includes aspects of athlete characteristics and behaviour, as well as opponent behaviour. Another description, with more emphasis on the biomechanical causes of the injury, might be that "the injury occurred as a result of a rapid sideways translation on a high friction surface, rotating while his foot remained firmly planted on the floor." A more detailed biomechanical description could be that "the injury occurred as a result of large external valgus moment and external rotation moment in combination with a translatory shift of the tibia relative to the femur." In other words, the description could include information ranging from the playing situation, player and opponent behaviour, to a more or less detailed biomechanical description of joint motion and loads.

An examination of the literature on sports injuries to see how the term "injury mechanism" is used to understand the mechanisms for ACL injuries shows that biomechanically oriented descriptions dominate, although to a different level of detail. Some studies only provide simple characteristics such as "contact/non-contact injuries" or "jumping/non-jumping injuries".

Others use terms like "side-step cutting manoeuvres", "tackle" or "long shot", "spiking" or "blocking", "phantom foot-mechanism"—descriptions that are related to a specific sport (European team handball, volleyball, and alpine skiing respectively). However, most of the more detailed studies use different biomechanical descriptors to depict the mechanism of injury. The level of detail varies here as well—for example, "deceleration injury" (describes whole body acceleration), "valgus torque" (describes knee kinetics), "anterior drawer" (describes the relative translation between femur and tibia), "quadriceps drawer" (describes the relative translation between femur and tibia as the result of quadriceps activation), and "inter-condylar lift off" (kinematic description of the result from, for example, a valgus or varus load).

It may be argued that, ultimately, the factors causing a joint injury are the loads that act on the joint (contact forces, ligament forces, and muscle forces), resulting from the external forces acting on the body and the internal forces and moments acting within the body. Therefore it is tempting to conclude that a detailed biomechanical description of these factors is the goal, and that this information is more important than a description of the playing situation. This may be a premature conclusion. We would argue that in many cases it is necessary to expand the traditional biomechanical approach to describing the inciting event if the objective is to prevent injuries. For example, an ankle sprain in football could be described as resulting from an inversion injury, a combined supination-internal rotation moment in plantar flexion that causes excessive loads to the anterior talofibular and calcaneofibular ligaments. However, if the injury results from illegal tackles from the side or behind, this information may be more relevant in a prevention perspective than a precise biomechanical description of ligament loading patterns. If severe sanctions to the player and team were introduced for such behaviour, this should result in fewer injuries.

Therefore we suggest that a full description of the mechanisms for a particular injury type in a given sport should include information on different levels. The description of the inciting event could be grouped into four categories, as shown in table. Hence, the term "injury mechanism" is here used to describe: (a) vital aspects of the playing (sports) situation—that is, the situation described from a sports specific point of view; (b) athlete and opponent behaviour—that is, a qualitative description of the athlete's action and interaction with the opponent; (c) gross biomechanical characteristics—that is, a description of whole body biomechanics; (d) detailed biomechanical characteristics—that is, a description of joint/tissue biomechanics.

		Non-contact ACL injury in basketball	**ACL injury in a mogul skiing jump landing**	**Knock-out in boxing**	**Lower leg stress fracture in football**
Playing (sports) situation	Team action	Fast break	Course steepness	Uppercut, hook	Exposure to matches and training (total load)
	Skill performed before, and at, the point of injury	Zone defence	Jump elements (e.g. twist, helicopter)	Counterattack	Midfielder with defensive and offensive tasks (i.e. many runs in matches/training)
	Court position	Charging	Jump height & length	Foot work	Hard working team
	Player position	Cutting		Forced into the corner/to the ropes	Frequency of duels
	Ball handling	Setting up for a shot		Ring-side referee decision	
		Defensive rebound		Inter-boxer distance	
		Man to man defence			
Athlete/ opponent behaviour	Player performance	Effort	Rhythm and balance before the jump	Awareness	Effort
	Opponent interaction	Disturbance by opponent	Concentration	Aggressiveness	Toe/heel runner
	Player attention	Intention	Balance	Punching power	Jumping technique
		Foot firmly fixed to the floor	Boot binding release	Punching speed	Duel technique

		Intention	Visual control	Balance	
		Technical foul	Jumping technique		
			Over-rotation		
			Falling technique		
Whole body biomechanics	Coarse description, often static, of whole body kinematics and kinetics	Sideways translation	Linear and angular momentum	Centre of mass velocity	Stride length
		Rotation of the body around the fixed foot	Energy absorption	Punching force	Stride frequency
		Speed at impact	Centre of mass to the rear	Punching direction	Vertical excursion
		Foot in front of centre of mass		Weight distribution on the legs	Ground reaction forces
					Knee flexion angle
Joint/tissue biomechanics	Detailed description of joint/tissue kinematics and kinetics	Valgus moment	Shear forces	Energy transfer	Bending moment
		Pivot shift of the tibia relative to the femur	Anterior drawer	Head acceleration	Shear forces
		Notch impingement	Intercondylar lift off, loading rate	Pressure distribution and localisation	Surface/shoe dynamics

9.5.3 Injury Model

In this model, intrinsic and extrinsic risk factors can affect load and load tolerance. For instance, the inter-condylar notch width has been discussed as a risk factor for non-contact ACL injuries in many studies. There are two major theories on how the notch width could influence injury rates. One is a possible association between notch width and ligament width (hence strength)—that is, ligament load tolerance as an intrinsic risk factor. The other hypothesis involves the interaction between an internal risk factor (that a narrow inter-condylar notch can lead to impingement of the ACL on the medial aspect of the femoral condyle) and the injury mechanism (that this may be more likely to occur when the knee is loaded in valgus and external rotation of the tibia). As mentioned above, it is even possible that this is more likely to occur on high friction floors (external risk factor) or to athletes with suboptimal neuromuscular control (internal risk factor).

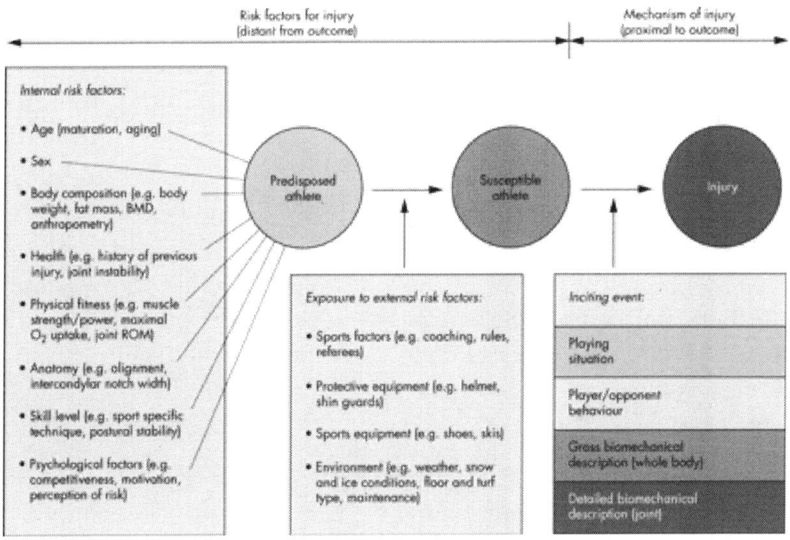

Fig Injury Model

The model can be used to study the interaction between different factors causing injury and address the potential for prevention—for example, for ankle sprains in volleyball or football. In both sports, the risk of an ankle sprain is 4–5 times higher if there is a history of previous injury to the same ankle (internal risk factor), mainly because of reduced neuromuscular function. Ankle sprains in volleyball mainly occur at the net when a player lands on the foot of an opponent or a teammate after blocking or attacking (injury mechanism). Ankle sprains in football mainly occur from late tackles from the side, often with impact by an opponent on the medial aspect of the leg just before or at foot strike, resulting in a laterally directed force (injury mechanism). These studies show that a significant proportion of ankle injuries are contact injuries resulting from a player landing on another player's foot (volleyball) or a medial blow to the ankle or lower leg (football), mechanisms where neither balance training nor ankle bracing would be expected to have a protective effect. However, it may be that the direct contact is not what causes the ligaments to tear, but merely serves to put the ankle in a vulnerable position when landing or running, especially in players with reduced neuromuscular control. Thus, on the basis of this, it may be hypothesised that increased neuromuscular control through training or bracing could help the player to position the foot correctly before putting weight on the ankle. Studies have shown that ankle sprains can be prevented through balance training and bracing in volleyball and football. In fact, the protective effects of taping and bracing have been persuasively shown in football, although only for players with previous ankle injury. In addition, in players with a history of previous injury, a balance training programme appears to reduce the risk of re-injury to the same level as healthy ankles in football and volleyball players. These findings illustrate how information on risk factors and injury mechanisms can be used to develop targeted prevention methods. This approach should be used for each specific injury type in a given sport.

9.6 Intrinsic and Extrinsic Factors Affecting Injury

9.6.1 Introduction

Both extrinsic and intrinsic factors can increase the risk of injury. Intrinsic factors include biomechanical deficiencies, including malignancy of limbs, muscular imbalances, degenerative processes, and other anatomical factors. Extrinsic factors include training errors, faulty technique, poor environmental conditions, incorrect equipment and surfaces. Extrinsic Factors of injury can be further classified into overuse, misuse, abuse, or disuse.

- Overuse is performing a task with a frequency that does not allow the tissues to recover and symptoms may be due to lack of muscle strength or endurance
- Misuse using improper form or equipment which may put abnormal stress on tissue structures.
- Abuse is having excessive force going through normal tissues.
- Disuse occurs after taking a period of time off without training resulting in deconditioning or altered neuromuscular control
- Excessive fatigue can negatively affect proper form and/or decreases the ability of stabilizers to maintain sound biomechanics. Careful monitoring of training volume, intensity, and duration will minimize overuse injuries.

Intrinsic factors such as stability and mobility deficiencies have been attributed to athletic and fitness related injuries. Compensatory movement patterns develop to overcome stability and mobility deficiencies. These poor movement patterns are used subconsciously whenever executing future tasks leading to greater mobility and stability imbalances and deficiencies which increases the risk of injury.

Prior injury is one of the most significant risk factors predisposing individuals to injuries. A decrease in proprioceptive input occurs if an injury is left untreated or inappropriately treated. A disruption in proprioception will negatively affect movement-pattern behaviors. The altered mobility, stability, and asymmetric influences eventually lead to compensatory movement patterns, increasing risk of injury. Risk of injury is increased under following conditions:

- Insufficient warm up
- Insufficient recovery
- Training stimulus is not progressive and regular
- Mechanical impairments

9.6.2 Biomechanical Impairments

Biomechanical impairments may contribute to orthopedic injury when combined with other factors than may negatively affect joint integrity (e.g. hyper-mobility, bio-structural weakness, insufficient adaptation of joint or muscular structures, fatigue, acute or chronic overtraining, etc.). For example, hypermobility does not produce joint instability unless the secondary stabilizers do not function adequately and symptoms occur.

- Weaknesses
- Flexibility
- Postural Impairments

The biomechanical impairments are listed for precautionary measures only, so they may be identified and possibly corrected in attempt to decrease risk of athletic injuries during exercise or physical activity. These impairments are possible risk factors for injury and may only increase injury when combined with other risk factors. A physician may need to establish if a biomechanical deficiency is structural, muscular, neuromuscular, or due to some other pathology.

The corrective exercises assume impairments are due to a muscular imbalance (flexibility, strength, or posture). Only a qualified physician or health care provider should diagnose and give prescription for an existing injury. In some circumstances, an attempt to correct a biomechanical impairment may irritate the injury and prolong recovery, particularly if certain therapy exercises are used inappropriately or initiated too soon or aggressively after an injury has occurred. The exercise selection should be based on positions that do not overstress the healing tissues. Even after underlying biomechanical impairments have been improved, a preexisting injury may require the attention of a physical therapist under the advice of a physician to restore total functionality.

9.6.3 Sports Conditioning

In athletic conditioning, improper planning and inadequate conditioning (both general and sports specific) prior to the competitive season are major causes of injury. In elite athletes, high volume and intensity of the training load is one of the major causes of injury.

An improperly planned conditioning program can be attributed to failure to understand and implement fundamental training principles and adaptation criteria through a periodized programming. A program that does not adequately prepare the athlete for the specific types of forces and stresses experienced on the field or court place athletes at risk.

Contrary to popular belief, conventional stretching per se does not appear to decrease the occurrence of injury. However, movement specific warm-ups before drills, tests, and the event performance can decrease risk of injury. Ideally, movement specific warm-ups should precede the event by several minutes, mimic the sports activity in the exact mechanics in which it will be performed, yet, not be so taxing as to compromise sports performance. Some sports are obviously more unpredictable than others and will not allow for rehearsal of every possible movement several minutes before they occur. Power or speed sports should consists a short series of movement specific warm-ups at progressive intensities. For example, several minutes before the start of 100m dash, a sprinter could perform a short progressive submax series of block starts immediately, followed by a short run, or simulated movement, thereby warming up in the same mechanics in which they will be engaging.

In the weight room, performing a movement specific warm-up set (e.g.: 50% 10RM) before workout sets, allow for performance benefits in addition to decreasing the risk of injury. Injuries can also be circumvented by well thought-out programs that adhere to evidence based research and sound training protocols that allow for adequate recovery. Also, using too much resistance or performing too many sets, particularly when athletes are first introduced to new movements or exercise variations are common training errors which can greatly increase the risk of injury.

The coach can introduce methods in which athletes can customized a 'group program' to each athlete's unique abilities, results, and needs. Coaches must allow for proper rehearsal and adaptation to 'new' exercises or movements, understand the benefits periodization techniques, adhere to adaptation criteria, implement program customization, use of movement specific -ups, and schedule adequate recovery periods for optimal progress.

References:

Williams JGP, Sperryn PN (eds): Sports Medicine, ed 2. Baltimore, Williams & Wilkins, 1976, pp 119123

Alley PW, Sells RL: Introduction to Physics. Boston, Allyn & Bacon, 1971

Madley R, Winter R: Modern Physics: A Student Study Guide. New York, John Wiley & Sons, 1971

Wilkie DR: Man as a source of mechanical power. Ergonomics, 3(1), 1960

Barham JN, Wooten EP: Structural Kinesiology. New York, Macmillan, 1973, pp 8586

Clauser CE, et al: Weight, Volume, and Center of Mass of Segments of the Human Body, Wright-Patterson Air Force Base, Ohio, AMRL-TR6970, 1969

Dumbleton JH, Black J: Principles of mechanics. In Black J, Dumbleton JH (eds): Clinical Biomechanics: Case History Approach. New York, Churchill Livingstone, 1981, pp 367-368

Clauser CE, McConville JT, Young JW: Weight, volume, and center of mass of segments of the human body. AMRL-TR6970, Wright-Patterson Air Force Base, Ohio, 1969

Schenck JM, Cordova FD: Introductory Biomechanics, ed 2. Philadelphia. F.A. Davis, 1980, pp 17.

Kendall HO, Kendall FP, Wadsworth GE: Muscles: Testing and Function, ed 2. Baltimore, Williams & Wilkins, 1971, p 18.

Squire PJ: Centre of gravity height and its relation to anthropometric measurements in female physical education students. Journal of Human Movements Studies, 3:214220, 1977.

Cochran GVB: A Primer of Orthopaedic Biomechanics. New York, Churchill Livingstone, 1982, pp 1415.

Barham JN, Wooten EP: Structural Kinesiology. New York, Macmillan, 1973, pp 87, 129130, 134135.

Gowitzke BA, Milner M: Understanding Scientific Basis of Human Movement, ed 2. Baltimore, Williams & Wilkins, 1980, p 55.

Williams JGP, Sperryn PN (eds): Sports Medicine, ed 2. Baltimore, Williams & Wilkins, 1976, pp 108110.

LeVeau B: Williams and Lissner: Biomechanics of Human Motion, ed 2. Philadelphia, W.B. Saunders, 1977, p 10.

Barham JN, Wooten EP: Structural Kinesiology. New York, Macmillan, 1973, pp 99100.

Edwards S: Physics: A Discovery Approach. New York, John Wiley & Sons, 1971.

Straley JW: Basic Physics. Englewood Cliffs, NJ, Prentice-Hall, 1974.

Cochran GVB: A Primer of Orthopaedic Biomechanics. New York, Churchill Livingstone, 1982, pp 1, 13, 16, 152153.

Straley JW: Basic Physics. Englewood Cliffs, NJ, Prentice-Hall, 1974.

Barham JN, Wooten EP: Structural Kinesiology. New York, Macmillan, 1973, p 131.

Schenck JM, Cordova FD: Introductory Biomechanics, ed 2. Philadelphia. F.A. Davis, 1980, pp 2224.

LeVeau B: Williams and Lissner: Biomechanics of Human Motion, ed 2. Philadelphia, W.B. Saunders, 1977, pp 46.

Dumbleton JH, Black J: Principles of mechanics. In Black J, Dumbleton JH (eds): Clinical Biomechanics: Case History Approach. New York, Churchill Livingstone, 1981, pp 360364.

White HE: Modern College Physics, ed 5. New York, Van Nostrand Reinhold, 1966.

Fowler RG, Myer DI: Physics for Engineers and Scientists. Boston, Allyn & Bacon, 1958.

Fuchs WR: Physics for the Modern Mind. New York, Macmillan, 1967.

Schenck JM, Cordova FD: Introductory Biomechanics, ed 2. Philadelphia. F.A. Davis, 1980, pp 79, 2526.

Gowitzke BA, Milner M: Understanding Scientific Basis of Human Movement, ed 2. Baltimore, Williams & Wilkins, 1980, p 78.

Black J, Dumbleton JH (eds): Clinical Biomechanics: Case History Approach. New York, Churchill Livingstone, 1981, pp 364365.

Cochran GVB: A Primer of Orthopaedic Biomechanics. New York, Churchill Livingstone, 1982, pp 5, 1217.

Williams JGP, Sperryn PN (eds): Sports Medicine, ed 2. Baltimore, Williams & Wilkins, 1976, pp 107108.

White AA, Panjabi MM: Clinical Biomechanics of the Spine. Philadelphia, J.B. Lippincott, 1978, pp 480481, 457.

Gonzna ER, Harrington IJ: Biomechanics of Musculoskeletal Injury. Baltimore, Williams & Wilkins, 1982, p 12.

Daniels L, Worthingham C: Therapeutic Exercise for Body Alignment and Function, ed 2. Philadelphia, W.B. Saunders, 1977, pp 101102.

LeVeau B: Williams and Lissner: Biomechanics of Human Motion, ed 2. Philadelphia, W.B. Saunders, 1977, pp 12.

Dumbleton JH, Black J: Principles of mechanics. In Black J, Dumbleton JH (eds): Clinical Biomechanics: Case History Approach. New York, Churchill Livingstone, 1981, pp 360361.

Hollingshead WH, Jenkins DB: Functional Anatomy of the Limbs and Back. Philadelphia, W.B. Saunders, 1981, pp 2829.

Strait LA, Inman VT, Ralston HJ: Sample illustrations of physical principles selected from physiology and medicine. American Journal of Physics, 15:375382, 1947.

White AA, Panjabi MM: Clinical Biomechanics of the Spine. Philadelphia, J.B. Lippincott, 1978, pp 330332.

Cailliet R: Low Back Pain Syndrome, ed 3. Philadelphia, F.A. Davis, 1981, pp 132137.

Karwowski W: Fuzzy modelling of stresses in manual lifting tasks. Ergonomics, 27:641649, 1984.

LeVeau B: Williams and Lissner: Biomechanics of Human Motion, ed 2. Philadelphia, W.B. Saunders, 1977, pp 2332.

Cochran GVB: A Primer of Orthopaedic Biomechanics. New York, Churchill Livingstone, 1982, pp 56, 5458.

Gowitzke BA, Milner M: Understanding Scientific Basis of Human Movement, ed 2. Baltimore, Williams & Wilkins, 1980, pp 4243.

Barham JN, Wooten EP: Structural Kinesiology. New York, Macmillan, 1973, pp 9193.

White AA III, Panjabi MM:

Spinal kinematics. In Goldstein M (ed): The Research Status of Spinal Manipulative Therapy, NINCDS Monograph No. 15, DHEW Publication No. (NIH) 76998, Stock No. 017-049-0 00607, Washington, DC, U.S. Government Printing Office, 1975, p 93.

LeVeau B: Williams and Lissner: Biomechanics of Human Motion, ed 2. Philadelphia, W.B. Saunders, 1977, pp 67.

Dumbleton JH, Black J: Principles of mechanics. In Black J, Dumbleton JH (eds): Clinical Biomechanics: Case History Approach. New York, Churchill Livingstone, 1981, pp 359360.

Barham JN, Wooten EP: Structural Kinesiology. New York, Macmillan, 1973, pp 100101.

Gowitzke BA, Milner M: Understanding Scientific Basis of Human Movement, ed 2. Baltimore, Williams & Wilkins, 1980, p 75.

McCormick WW: Fundamentals of College Physics. New York, Macmillan, 1965.

Shames IH: Engineering Mechanics--Statics and Dynamics, ed 2. Englewood Cliffs, NJ, Prentice-Hall, 1967.

Dempster WT: Free-body diagrams as an approach to the mechanics of human posture and locomotion. In Evans FG (ed): Biomechanical Studies of the MusculoskeletaL System. Springfield, IL, Charles C. Thomas, 1961.

Dumbleton JH, Black J: Principles of mechanics. In Black J, Dumbleton JH (eds): Clinical Biomechanics: Case History Approach. New York, Churchill Livingstone, 1981, pp 368369.

White AA, Panjabi MM: Clinical Biomechanics of the Spine. Philadelphia, J.B. Lippincott, 1978, pp 476*477,

Dumbleton JH, Black J: Principles of mechanics. In Black J, Dumbleton JH (eds): Clinical Biomechanics: Case History Approach. New York, Churchill Livingstone, 1981, pp 365-367.

Schenck JM, Cordova FD: Introductory Biomechanics, ed 2. Philadelphia. F.A. Davis, 1980, pp 7681, 85.

White AA III, Panjabi MM: Spinal kinematics. In Goldstein M (ed): The Research Status of Spinal Manipulative Therapy, NINCDS Monograph No. 15, DHEW Publication No. (NIH) 76998, Stock No. 017-049-0 00607, Washington, DC, U.S. Government Printing Office, 1975, p 93.

Sears FW, Zemansky MW: University Physics, ed 4. Reading, MA, Addison- Wesley, 1960.

Halliday D, Resnick R: Fundamentals of Physics. New York, John Wiley & Sons, 1974.

LeVeau B: Williams and Lissner: Biomechanics of Human Motion, ed 2. Philadelphia, W.B. Saunders, 1977, p 9.

White AA, Panjabi MM: Clinical Biomechanics of the Spine. Philadelphia, J.B. Lippincott, 1978, p 462.

White AA, Panjabi MM: Clinical Biomechanics of the Spine. Philadelphia, J.B. Lippincott, 1978, p 504.

Jokl P: Muscle. In Albright JA, Brand RA (eds): The Scientific Basis of Orthopaedics. New York, Appleton-Century-Crofts, 1979, pp 380381.

Karvinen E, Komi PV: Neuromuscular performance. In Larson LA (ed): Fitness, Health, and Work Capacity. New York, Macmillan, 1974, pp 8586.

Morehouse LE, Miller AT Jr: Physiology of Exercise. St. Louis, C.V. Mosby, 1948, pp 2122.

Dumbleton JH, Black J: Principles of mechanics. In Black J, Dumbleton JH (eds): Clinical Biomechanics: Case History Approach. New York, Churchill Livingstone, 1981, p 387.

Lehman RL, Swartz C: Foundations of Physics. New York, Holt, Rinehart & Winston, 1965.

LeVeau B: Williams and Lissner: Biomechanics of Human Motion, ed 2. Philadelphia, W.B. Saunders, 1977, pp 1516.

Schenck JM, Cordova FD: Introductory Biomechanics, ed 2. Philadelphia. F.A. Davis, 1980, pp 7681, 83.

Benedek GB, Villars FMH: Physics. Reading, PA, Addison-Wesley, 1973.

Kenedi RM (ed): Biomechanics and Related Bioengineering Topics. Oxford, Pergamon Press, 1965.

Sayers BMcA, et al: Engineering in Medicine. London, Oxford University Press, 1974.

Morehouse LE, Miller AT Jr:

Physiology of Exercise. St. Louis, C.V. Mosby, 1948, pp 224-225.

Hollingshead WH, Jenkins DB: Functional Anatomy of the Limbs and Back.Philadelphia, W.B. Saunders, 1981, pp 3436.

Williams JGP, Sperryn PN (eds): Sports Medicine, ed 2. Baltimore, Williams & Wilkins, 1976, pp 114-118.

Gonzna ER, Harrington IJ: Biomechanics of Musculoskeletal Injury. Baltimore, Williams & Wilkins, 1982, pp 34-35.

Morehouse LE, Cooper JM: Kinesiology. St. Louis, C.V. Mosby, 1950, pp 101-102

Morehouse LE, Miller AT Jr: Physiology of Exercise. St. Louis, C.V. Mosby, 1948, pp 223-224.

Cochran GVB: A Primer of Orthopaedic Biomechanics. New York, Churchill Livingstone, 1982, pp 18, 225-227.

Morehouse LE, Cooper JM: Kinesiology. St. Louis, C.V. Mosby, 1950, pp 103-106.

Gowitzke BA, Milner M: Understanding Scientific Basis of Human Movement, ed 2. Baltimore, Williams & Wilkins, 1980, pp 44-47.

Barham JN, Wooten EP: Structural Kinesiology. New York, Macmillan, 1973, pp 82-83.

LeVeau B: Williams and Lissner: Biomechanics of Human Motion, ed 2. Philadelphia, W.B. Saunders, 1977, pp 41-42.

Cochran GVB: A Primer of Orthopaedic Biomechanics. New York, Churchill Livingstone, 1982, pp 19-20.

White AA III, Panjabi MM: Spinal kinematics. In Goldstein M (ed): The Research Status of Spinal Manipulative Therapy, NINCDS Monograph No. 15, DHEW Publication No. (NIH) 76998, Stock No. 017-049-0 00607, Washington, DC, U.S. Government Printing Office, 1975, p 98.

White AA, Panjabi MM: Clinical Biomechanics of the Spine. Philadelphia, J.B. Lippincott, 1978, pp 457-458, 476.

Gonzna ER, Harrington IJ: Biomechanics of Musculoskeletal Injury. Baltimore, Williams & Wilkins, 1982, pp 37-38.

Dumbleton JH, Black J: Principles of mechanics. In Black J, Dumbleton JH (eds): Clinical Biomechanics: Case History Approach. New York, Churchill Livingstone, 1981, pp 374-376, 394, 398.

Williams JGP, Sperryn PN (eds): Sports Medicine, ed 2. Baltimore, Williams & Wilkins, 1976, pp 125-126

Gowitzke BA, Milner M: Understanding Scientific Basis of Human Movement, ed 2. Baltimore, Williams & Wilkins, 1980, pp 73, 77.

White AA, Panjabi MM: Clinical Biomechanics of the Spine. Philadelphia, J.B. Lippincott, 1978, p 506.

Morehouse LE, Cooper JM: Kinesiology. St. Louis, C.V. Mosby, 1950, pp 109-110.

White AA, Panjabi MM: Clinical Biomechanics of the Spine. Philadelphia, J.B. Lippincott,

Terms

Absolute – absolute most commonly means that the value being reported is not being expressed relative to bodyweight.

Acceleration – acceleration is the rate of change of velocity of an athlete (in m/s/s), which is the change in velocity divided by time. It is also vector quantity (meaning that it has a direction associated with it) and is proportional to both the external force exerted upon it and its mass.

Accommodating resistance – strictly, accommodating resistance is the same as isokinetic resistance, where the load "accommodates" the force applied by the subject so that velocity remains constant throughout the movement. In practice, the term "accommodating resistance" is often used to refer to bands and chains, which are a form of variable resistance that approximate isokinetic external resistance.

Active – active refers to circumstances in human movement where movements or moments are produced by muscle forces that are generated by muscle contractions following from muscle activity.

Anatomical cross-sectional area – the area of the cross section of a muscle perpendicular to its longitudinal axis.

Anterior – situated on or closer to the front aspect of the body.

Center of mass (COM) – the center of mass is the unique point within an object where the weighted relative position of the distributed mass of the object sums to zero. In biomechanics, it is common to assume that an object behaves as if it were a point mass located at the COM rather than a distributed mass, for ease of calculations.

Concentric – concentric muscle actions occur when muscles shorten under tension, leading to changes in joint position.

Displacement – displacement is a change in position of an object. Jumping height is an example of a displacement that is often measured in biomechanics.

Distal – situated further away from the center of the body.

Eccentric – eccentric muscle actions occur when muscles lengthen under tension, leading to changes in joint position.

External joint moments – Resistance in strength training produces an external moment, whereas

Fascicle length – the length of an individual muscle fascicle, where a muscle fascicle is a bundle of muscle fibers

Fiber length – the length of an individual muscle fiber, noting that an individual muscle fiber does not generally run the entire length of a muscle fascicle

Flight time – flight times are the durations of time in which an athlete is not in contact with the ground during athletic movements, such as during vertical jumps or sprint running. During vertical jumps, flight time can be used to estimate jump height by using Newton's laws of motion.

Ground contact time – ground contact times are the durations of time in which feet are in contact with the ground during athletic movements, such as drop jumps or sprint running. Sprint running displays some of the shortest ground contact times, which are around 0.1 seconds, which does not allow much time for the athletes to exert force to propel themselves forwards.

Impulse – impulse is force multiplied by time over which the force acts, expressed as Ns.

Inferior – situated on or closer to the underneath of the body.

Internal joint moments – muscles produce an internal moment to counteract the external moment. Moments are usually measured in Newton-meters.

Isoinertial resistance – isoinertial external resistance is any external resistance that remains constant throughout the movement. Most typical free weight exercises are isoinertial, as the mass or load do not change during the exercise.

Isokinetic resistance – isokinetic external resistance is a form of variable resistance that maintains the velocity of the movement constant irrespective of the force applied. It is only achievable in practice using a dynamometer.

Isometric – isometric muscle actions occur when no movement in the joint take place. The muscle does shorten and the tendon does lengthen but this does not alter joint position.

Joint angular velocity – joint angular velocity is the rate of change of joint ROM, which is therefore measured in degrees/s or rad/s.

Joint moment – a moment is the turning effect produced by a force. It is often synonymous with torque, which can be though of as the rotational analog to linear force (turning force), and is calculated by multiplying the perpendicular force by the distance from the pivot (or axis of rotation).

Joint power – it is possible to measure the power output of individual joints during movement by multiplying the torque by the joint angular velocity. It is usually reported in Newton-meters per second.

Lateral – situated closer to the sides of the body, or away from the midline of the body.

Medial – situated further from the sides of the body, or closer to the midline of the body.

Momentum – momentum is mass multiplied by velocity, expressed in kg m/s.

Muscle architecture – Muscle architecture is the organization of the muscle fibers within the muscle relative to the line of force generation. It comprises three main aspects: normalized fiber length, pennation angle, and physiological cross-sectional area.

Normalized fiber length – the length of an individual muscle fiber, normalized for its average sarcomere length.

Passive – passive refers to circumstances in human movement where movements or moments are produced by muscle forces that are not generated by muscle contractions following from muscle activity but are instead are generated by the elastic properties of the muscle fibers themselves, as well as of the ligaments, bones, tendons, and fascia.

Pennation angle – this is the angle between the line of muscle fiber alignment and the longitudinal axis of the muscle.

Physiological cross-sectional area – the area of the cross section of a muscle perpendicular to the direction of its fibers, generally at its largest point.

Posterior – situated on or closer to the rear aspect of the body.

Power – power is the rate at which work is done, and can be calculated either by dividing the work done by the time in which the work was done or by multiplying the force applied by the velocity at which it was applied, expressed in Watts (W).

Proximal – situated closer towards the center of the body.

Range of motion (ROM) – the displacement in angular movement is generally measured by reference to ROM, which can be reported in either degrees or radians (rad).

Rate of force development (RFD) – RFD is the rate of change of force over time, expressed as N/s.

Rate of torque development (RTD) – and is usually measured in Newton-meters per second.

Regional hypertrophy – this is the non-uniform increase in cross-sectional area along the length or across the breadth of a muscle.

Relative – relative most commonly means expressing another biomechanical value in relation to bodyweight. It can be calculated by dividing a value by the bodyweight of the subjects. Where mean relative values are presented, such as joint moments, these can be converted to absolute values by multiplying the relative value by the mean bodyweight of the subjects in the study.

Superior – situated on or closer to the top of the body.

Variable resistance – variable external resistance is a form of external resistance that changes throughout the movement. This can be achieved in practice using bands and chains in combination with barbells or by using machine weights with cam devices.

Velocity – velocity is the rate of change of position of an athlete (in m/s), which is the displacement divided by time. It is a vector quantity (meaning that it has a direction associated with it) and speed is its scalar equivalent.

Work – work is force multiplied by the distance moved as a result of the force acting, expressed in Joules (J).

Index